Ceramic Electrolytes for All-Solid-State Li Batteries

Related Books from World Scientific

Handbook of Solid State Batteries
Second Edition
edited by Nancy J Dudney, William C West and Jagjit Nanda
ISBN: 978-981-4651-89-9

Prospects for Li-ion Batteries and Emerging Energy Electrochemical
Systems
edited by Laure Monconduit and Laurence Croguennec
ISBN: 978-981-3228-13-9

Related Journal from World Scientific

Functional Materials Letters
Editor-in-Chief: Li Lu
Print ISSN: 1793-6047
Online ISSN: 1793-7213

Ceramic Electrolytes for All-Solid-State Li Batteries

Masashi Kotobuki
National University of Singapore, Singapore

Shufeng Song
Chongqing University, China

Chao Chen
National University of Singapore, Singapore
National University of Singapore Suzhou Research Institute, China

Li Lu
National University of Singapore, Singapore
National University of Singapore Suzhou Research Institute, China

World Scientific

NEW JERSEY · LONDON · SINGAPORE · BEIJING · SHANGHAI · HONG KONG · TAIPEI · CHENNAI · TOKYO

Published by

World Scientific Publishing Co. Pte. Ltd.

5 Toh Tuck Link, Singapore 596224

USA office: 27 Warren Street, Suite 401-402, Hackensack, NJ 07601

UK office: 57 Shelton Street, Covent Garden, London WC2H 9HE

Library of Congress Cataloging-in-Publication Data

Names: Kotobuki, Masashi, 1974–   author.

Title: Ceramic electrolytes for all-solid-state Li batteries / by Masashi Kotobuki
(National University of Singapore, Singapore) [and three others].

Other titles: Ceramic electrolytes for all-solid-state lithium ion batteries

Description: New Jersey : World Scientific, 2018. | Includes bibliographical references.

Identifiers: LCCN 2018000456 | ISBN 9789813233881 (hardcover : alk. paper)

Subjects: LCSH: Lithium ion batteries. | Solid state batteries. | Electronic ceramics. |
Electrolytes--Conductivity.

Classification: LCC TK2945.L42 C47 2018 | DDC 621.31/2424--dc23

LC record available at https://lccn.loc.gov/2018000456

British Library Cataloguing-in-Publication Data

A catalogue record for this book is available from the British Library.

For any available supplementary material, please visit
http://www.worldscientific.com/worldscibooks/10.1142/10815#t=suppl

Typeset by Stallion Press
Email: enquiries@stallionpress.com

Preface

Li-ion batteries have contributed to rapid development and growth of portable devices such as cellular phones and laptop computers due mainly to their high portability since their commercialization in 1991. Recently, Li-ion batteries have also successfully been used in large format energy storages such as electrical vehicles and grade balancing. Although Li-ion batteries have already been widely used in many different areas, they are still facing a lot of issues including poor safety, short performance life, and relatively low specific energy. To address those issues, new format of batteries, namely solid-state and all-solid-state Li batteries, have been developed. Solid-state and all-solid-state Li batteries have been recognized to be suitable formats as next-generation energy storage devices due to high energy density, high safety and high durability. Therefore, intensive efforts have been devoted to develop these types of Li batteries, in particularly to develop Li ion conductive ceramics, also named as ceramic electrolytes.

This book provides principles and fundamentals of ion conduction mechanisms of ion conductors, recent research progress of ceramic electrolytes and their applications in all-solid-state Li batteries.

This book is composed of eight chapters. After a brief introduction of ceramic electrolytes in Chapter 1, a detailed description about current Li-ion battery system is shown in Chapter 2, where materials and their properties of cathode, anode and electrolytes are described minutely. Chapter 3 presents the history of solid electrolyte since the discovery of ion conduction in ceramics in 1833. Chapter 4 dedicates to the understanding of ion conduction mechanisms. Ion conduction mechanisms inside crystal (hopping model) and at grain boundaries (space charge and percolation effects) are explained in detail. Chapters 5 and 6 focus on structures, properties and features of various kinds of ceramic electrolytes. Application of the ceramic electrolytes in the all-solid-state batteries is described in Chapter 7. Four types of the all-solid-state batteries, i.e. thin film form, bulk form, Li-S and Li-air batteries, are thoroughly described from their principle to recent research progress. Finally, current issues in the R&D on the all-solid-state batteries are explained.

This book covers a wide range of ceramic electrolytes and includes useful information for development of solid electrolytes. We truly believe that this book will be very useful not only for researchers but also for engineers for development of next-generation energy storage devices.

December, 2017

Contents

Chapter 1

Introduction

A ceramic material is an inorganic and non-metallic solid material that is composed of metal, non-metal and metalloid atoms bound by covalent or ionic bonds. The ceramic materials can range from crystalline and semi-crystalline to amorphous (glass) depending on the ordering of the structure. Generally, the ceramic materials have a high melting temperature, poor thermal and electrical conductivities, high hardness and low ductility. Also, the ceramic materials are lighter (less dense) than metals and heavier than polymers. They normally possess high thermal resistance, but low resistance for heat shock. Because most elements and types of bonding, and all range of crystallinity can comprise the ceramics materials, a lot of materials such as oxide, nitride and carbide materials are included in this category. Accordingly, the ceramic materials have been employed in a wide range of applications, such as in the manufacturing of bricks, tiles, dishes and vases.

The ceramic materials have been widely used in our daily lives since the ancient times. The earliest ceramics made by human are thought to be pottery objects. It was made of clay, either alone or mixed with other materials like sand (silica), followed by hardening and sintering by fire. Later, glazes were discovered and developed. They are normally glassy and amorphous ceramics with low melting point, and they can decrease the porosity of crystalline ceramic surface once they are coated on them. Because of the glazes, a smooth and colored ceramic surface can be obtained, and so a wide range of ceramic art was developed. Ceramics now include domestic, industrial and building products as well as ceramic art. In the 20th century, new ceramics materials such as semiconductors appeared in advanced ceramic engineering [1]. Particularly, developing ceramics parts for gas turbine engines are expected to reduce the volume and weight of the cooling system and even limit operating temperatures due to the high thermal stability of the ceramic materials. Also, bioceramics, such as those used in dental implants and synthetic bones, have played an important role in the medical field. Other examples of bioceramics are pacemakers, kidney dialysis machines and respirators. Calcium phosphate-based ceramics have been used for bone substitution. These ceramics have a similar chemical composition and structure as the bones. In dental implants, a polymer–ceramic composite is used to fill the pores of ceramics. The high biocompatibility is strongly required for the bioceramics.

The properties of ceramic materials are a direct result of its crystal structure and chemical composition. In most ceramic materials, ions cannot migrate because all ions must form a rigid skeleton structure. However, in some ceramics, only one ion can move with a low energy barrier and as a result, these ceramics possess high ionic conductivities comparable to those of molten salts and liquid electrolytes. Such ceramics are called "ion conductive ceramics." The ion conductive ceramics have gained much attention for a wide variety of electrochemical applications, ranging from power generation (e.g. solid oxide fuel cell, SOFC)and energy storage (e.g. battery and capacitor) to atomic switches [2], and intensively researched to improve their properties, especially ionic conductivity, and to clarify the origin of their high ionic conductivities. The ion conductive ceramics are also called "solid electrolytes" when they are applied in electrochemical devices. In particular, the application of ion conductive ceramics to energy storage devices has been prompted in many groups.

Nowadays, we are facing global warming problem due to emission of CO_2 as a result of burning fossil fuels like gasoline. In order to build a sustainable society and use renewable energy, usage of natural energy sources such as wind and solar powers has been researched.

However, the intermittent nature of natural energy sources precludes it from being a stable energy supply. Therefore, the development of energy storage devices is key to enable the usage of natural energy.

There are two kinds of electric energy storage devices, i.e. capacitors and batteries. The difference between these two devices is in the storage mechanism of electric charge. The capacitors store electric charges in the vicinity of the surface of the electrode (electrochemical double layer) (Fig. 1.1), while the batteries store the electric charges inside the electrode by Faradic reactions (Fig. 1.2). Therefore, the battery usually has higher capacity than the capacitors because

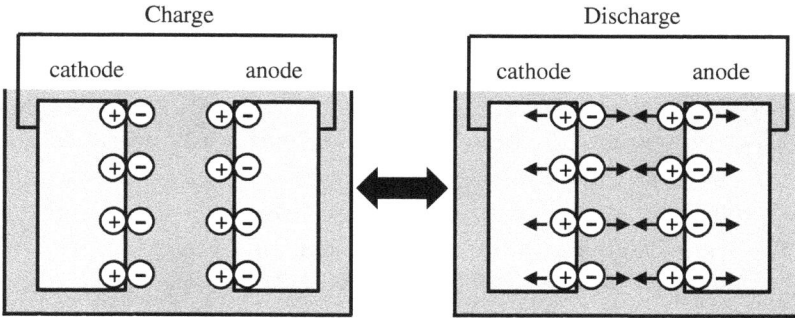

Fig. 1.1 Structure of a capacitor.

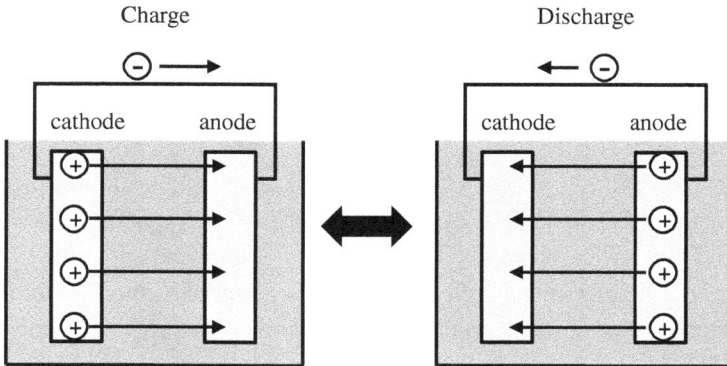

Fig. 1.2 Structure of battery.

the whole electrode can store the electric charges. On the contrary, the capacitors can charge and discharge faster than the batteries because they are not hindered by the kinetics of the Faradic reaction. The capacity of a capacitor is strongly influenced by the surface of the electrodes where electric charges are stored. Active carbon, which possesses a high surface area, has been used for the electrode material of a capacitor for this reason. However, electrode materials of batteries depend on battery reaction. For example, different electrode materials are used in Ni-hydrogen and lead-acid batteries.

Some Faradic reactions are very fast. The storage devices using the fast Faradic reactions are also battery, but they behave like capacitors. Such capacitors are called pseudo-capacitors. In the pseudo-capacitors, the electric charges are thought to be stored inside the electrode, but they are probably stored only in the vicinity of the surface of the electrodes.

Electrolytes are put between cathode and anode to avoid short circuit. Normally, liquid electrolytes have been used in batteries. The liquid electrolyte sometimes causes serious safety issues like electrolyte leakage and evaporation. Especially, energy storage from natural energy resources requires a large-scale battery which contains more electrolytes, and thus the electrolyte problems also become more serious. Therefore, the development of solid electrolytes (ion conductive ceramics) has been intensively researched.

Among the innumerable researches on the conductive ceramics, many ion species such as O^{2-} [3], Li^+ [4–6], Ag^+ [7], Na^+ [8], Cu^{2+} [9], Mg^{2+} [10], Al^{3+} [11], Sn^{4+} [12], H^+ [13] and so on have been found to migrate in ceramics depending on the structure of the ceramics and the temperature. Structures of fluorite-, perovskite-, apatite-, K_2NiF_4-type and AgI are well known as the ion conductive ceramics [14] (Fig. 1.3). AgI and $Rb_4Cu_{16}I_7Cl_{13}$ were thought to be suitable solid electrolytes for electrochemical devices due to their high ionic conductivity; however, their cost and difficulty in preparation have restricted their applications [15]. Alternatively, O^{2-}, Na^+ and Li^+ conductive ceramics have been given more attention for application in electrochemical devices due to their high

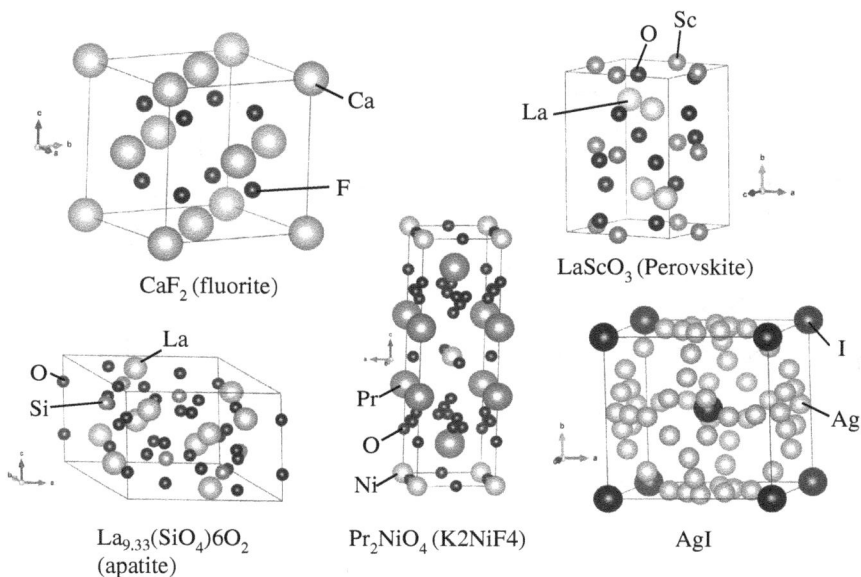

Fig. 1.3 Structure of solid electrolytes.

ion conductivities, low cost of production and wide electrochemical windows. The O^{2-} conductive ceramics such as Yttria-stabilized zirconia (YSZ) can be used as a solid electrolyte in SOFC, which is expected to be employed in co-generation system and fuel-cell vehicles. In the SOFC, oxygen is supplied to the cathode and oxygen ion moves to the anode through the solid electrolyte. The oxygen ions produce water after reacting with hydrogen at the anode, as shown in Fig. 1.4(a). Usually, (La, Sr)MnO$_3$ or LaSr(Co, Fe)O$_3$ and Ni–YSZ cermet are used at the cathode and anode, respectively. This system can produce electricity with emission of only H_2O, not CO_2, and hence can be expected to be a clean power generation system. On the other hand, Na$^+$ conductive ceramics such as β-Al$_2$O$_3$ solid electrolyte were successfully developed for the Na–S battery, which has been already commercialized. Na metal and sulfur are used as anode and cathode, respectively (Fig. 1.4(b)). Na ions migrate from the anode to the cathode through the β-Al$_2$O$_3$ solid electrolyte during discharging. At the cathode, sulfur reacts with Na and forms

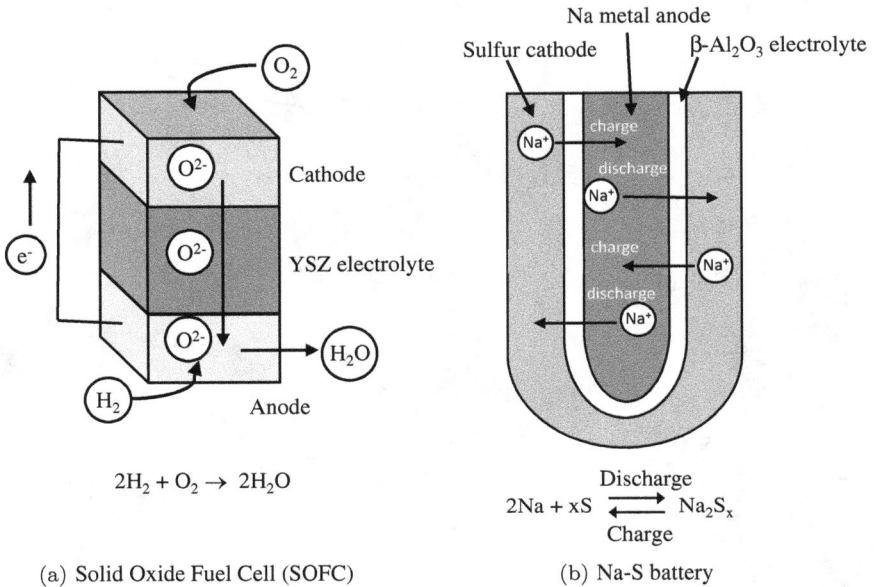

(a) Solid Oxide Fuel Cell (SOFC)

$$2H_2 + O_2 \rightarrow 2H_2O$$

(b) Na-S battery

$$2Na + xS \underset{Charge}{\overset{Discharge}{\rightleftharpoons}} Na_2S_x$$

Fig. 1.4   Structure of (a) SOFC and (b) Na-S battery.

a Na–S alloy. This system possesses a high energy density — about three times higher than that delivered by a conventional lead-acid battery — long life and low cost(all components are abundant). This system is being used in factories and power plants for load-levelling and emergency power supply. Both systems do not contain any liquid components. Therefore, some issues like electrolyte leakage and evaporation, which are sometimes caused by the liquid electrolyte in the conventional commercial batteries such as nickel-hydrogen, Li and lead-acid batteries, never occur in these systems.

The currently available commercial Li batteries contain flammable organic liquid electrolyte. The flammability of the electrolyte is always a concern with regard to safety of the Li batteries as well as electrolyte leakage and evaporation, particularly in the large-scale development of Li batteries for electric vehicle (EV) and hybrid vehicles (HV) and for stationary energy storage. Li-ion conductive ceramics have been anticipated as a solid electrolyte for Li batteries (all-solid-state Li batteries) due particularly to their non-flammability. The development of Li-ion conductive ceramics as often named as solid electrolytes is a key issue for realization of the all-solid-state Li batteries as present electrode materials are thought to be used for the all-solid-state batteries. Therefore, many researchers have been vigorously searching new solid electrolytes, identifying the structure of them, characterizing Li-ion conductivity and studying Li-ion conduction path in them. The solid electrolytes for the Li batteries can be divided into three groups, i.e. crystalline group, amorphous/glass group and glass-ceramics group [16], each of which has its advantages and disadvantages in terms of conductivity, processability, durability and cost.

References

[1] Wikipedia, https://en.wikipedia.org/wiki/Ceramic.

[2] T. Hasegawa, K. Terabe, T. Sakamoto, M. Aono: Nanoionics switching devices: "Atomic switches", *MRS Bull.* 34 (2009) 929–934.

[3] N. Mahato, A. Banerjee, A. Gupta, S. Omar, K. Balani: Progress in material selection for solid oxide fuel cell technology: A review, *Prog. Mater. Sci.* 72 (2015) 141–337.

[4] M. Kotobuki, M. Koishi, Y. Kato: Preparation of $Li_{1.5}Al_{0.5}Ti_{1.5}(PO_4)_3$ solid electrolyte via a co-precipitation method, *Ionics* 19 (2013) 1945–1948.

[5] M. Kotobuki, M. Koishi: Sol–gel synthesis of $Li_{1.5}Al_{0.5}Ge_{1.5}(PO_4)_3$ solid electrolyte, *Ceram. Intl.* 41 (2015) 8562–8567.

[6] M. Kotobuki, M. Koishi: Preparation of $Li_7La_3Zr_2O_{12}$ solid electrolyte via a sol-gel method, *Ceram. Intl.* 40 (2014) 5043–5047.

[7] S. Hull, D. A. Keen, D. S. Sivia, P. A. Madden, M. Wilson: The high-temperature superionic behaviour of Ag_2S, *J. Phys.: Condensed Matter* 14 (2002) L9–L17.

[8] B. H. Karina, M. Armand, T. Rojo: High temperature sodium batteries: status, challenges and future trends, *Energy Environ. Sci.* 6 (2013) 734–749.

[9] C. Shi, X. Xi, Z. Hou, E. Liu, W. Wang, S. Jin, Y. Wu, G. Wu: Atomic-level characterization of dynamics of copper ions in CuAgSe, *J. Phys. Chem. C* 120 (2016) 3229–3234.

[10] S. Tamura, M. Yamae, Y. Hoshino, N. Imanaka: Highly conducting divalent Mg^{2+} cation solid electrolytes with well-ordered three-dimensional network structure, *J. Solid State Chem.* 235 (2016) 7–11.

[11] N. Imanaka, S. Tamura, M. Hiraiwa, G. Adachi: trivalent aluminum ion conducting characteristics in $Al_2(WO_4)_3$ single crystals, *Chem. Mater.* 10 (1998) 2542–2545.

[12] N. Nunotani, T. Ohsaka, S. Tamura, N. Imanaka: Tetravalent Sn^{4+} ion conductor based on NASICON-type phosphate, *ECS Electrochem. Lett.* 1(4) (2012) A66–A69.

[13] M. Dippon, S. M. Babinies, H. Ding, S. Ricote, N. P. Sullivan: Exploring electronic conduction through $BaCe_xZr_{0.9-x}Y_{0.1}O_{3d}$ proton-conducting ceramics, *Solid State Ionics* **286** (2016) 117–121.

[14] "Ion conductive solid", Toray Research Center (2012).

[15] J. Gao, Y-.S. Zhao, S.-Q. Shi, H. Li: Lithium-ion transport in inorganic solid electrolyte, *Chin. Phys. B* **25** (2016) 018211.

[16] Y. Ren, K. Chen, R. Chen, T, Liu, Y. Zhang, C.-W. Nan: Oxide electrolytes for lithium batteries, *J. Am. Ceram. Soc.* **98** (2015) 3603–3623.

Chapter 2

Li Battery

Recent rapid progress and growing efforts to use clean and renewable energies have strongly required to urgently develop energy storage devices with high energy and power densities [1]. Electrochemical energy storage devices such as batteries and capacitors match this requirement because they can efficiently store and deliver energy when it is needed. Particularly, Li battery is expected to possess high energy density compared with other batteries like lead-acid battery, and so it has been thought to be a promising battery to fulfill this requirement [2].

Table 2.1 summarizes energy density and operation voltage of various commercial rechargeable batteries currently available [3]. The energy density of Li batteries is the highest among them. The energy density is calculated by multiplying the capacity (Ah kg^{-1}) with the operation voltage (V) of the battery. As can be seen in Table 2.1, the operation voltage of the Li battery is much higher than those of other batteries. In other words, the high operation voltage is the reason for the high energy density of Li batteries. Due to the high energy density, Li-ion batteries have triggered the growth of small portable electronic devices such as laptop computers and mobile phones since it was first commercialized in 1991 [4]. Today, Li-ion batteries are produced at the rate of billions of units per year.

Figure 2.1 depicts the most conventional structure of Li-ion battery using a graphite anode and a LiCoO$_2$ cathode. This structure

Table 2.1   Energy density of various commercialized batteries.

Battery	Practical energy density (Wh kg^{-1})	Theoretical energy density (Wh kg^{-1})	Operation voltage (V)
Li battery	110–160	400	3.8
Lead-acid battery	30–50	170	2.0
Nickel-cadmium battery	45–80	245	1.3
Nickel-hydrogen battery	60–120	280	1.4

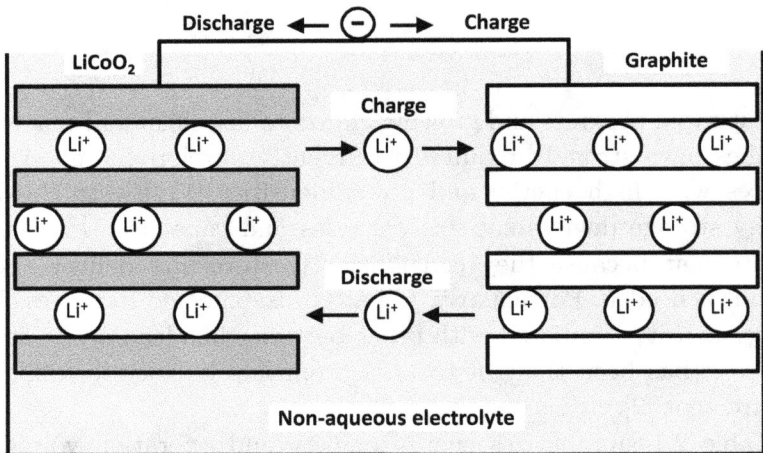

Fig. 2.1   Structure of commercial Li-ion battery.

was employed in the first commercial Li-ion batteries. The Li batteries do not contain Li-metal as anode. Li ion is supplied only by the cathode. Therefore, the Li batteries are also called Li-ion batteries. The Li-ion batteries seem to be a kind of Li-ion device and include a non-aqueous electrolyte and a Li intercalated cathode and anode. During discharge, the Li ions move from the graphite anode to the LiCoO$_2$ cathode. While charging, the Li ions migrate in the opposite direction. The battery reaction of the Li-ion batteries (using graphite anode and LiCoO$_2$ cathode) is as follows:

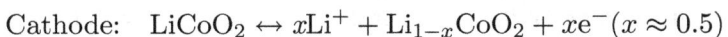

$$\text{Cathode:}\quad LiCoO_2 \leftrightarrow xLi^+ + Li_{1-x}CoO_2 + xe^- \ (x \approx 0.5)$$

Anode: $6C + xLi^+ + xe^- \leftrightarrow Li_xC_6$

Overall: $LiCoO_2 + 6C \leftrightarrow Li_{1-x}CoO_2 + Li_xC_6$

As can be seen, the electrolyte does not appear in the battery reaction since it works only as a Li-ion conducting media. Owing to this feature, the structure of the Li-ion batteries is also called "Rock-chair structure". Fewer amounts of electrolytes can reduce diffusion length of the Li ion and the weight and volume of the battery, resulting in improved performance of the Li-ion batteries. This feature is different from lead-acid batteries which have been used in the car for a long time. In the lead-acid battery, Pb ions react with H_2O and SO_4^{2-} ions in the electrolyte. The battery reaction of lead-acid batteries is as follows:

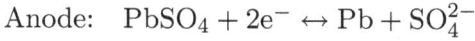

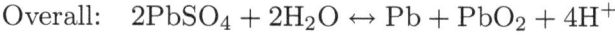

Cathode: $PbSO_4 + 2H_2O \leftrightarrow PbO_2 + 4H^+ + SO_4^{2-} + 2e^-$

Anode: $PbSO_4 + 2e^- \leftrightarrow Pb + SO_4^{2-}$

Overall: $2PbSO_4 + 2H_2O \leftrightarrow Pb + PbO_2 + 4H^+$

In the lead-acid batteries, a certain amount of electrolyte is needed to complete the battery reaction. This increases the weight of battery and reduces energy density. On the contrary, the amount of electrolyte can be reduced without limitation by the battery reaction in the Li batteries.

Electrodes of Li batteries are generally made of three components, namely, active materials which are involved in the battery reaction, conductive material and binder which help to maintain electron and ion conductive paths. Carbon materials such as Ketjen black and acetylene black are used as the conductive materials. As the binder, PVdF (polyvinylidene fluoride) has been exclusively used so far. These three components are mixed with a solvent (*N*-methyl-2-pyrrolidone, NMP). Owing to the toxicity of NMP, new binders which can be dissolved in water [5] as well as binder-free electrodes [6] have also been studied. However, the PVdF-NMP-based binder system is still widely used. The slurry obtained, containing the electrode material, carbon material, PVdF and NMP, is painted on a current collector. After drying, it is employed as an electrode. The thickness of the electrode is generally below 100 μm (Fig. 2.2). The current

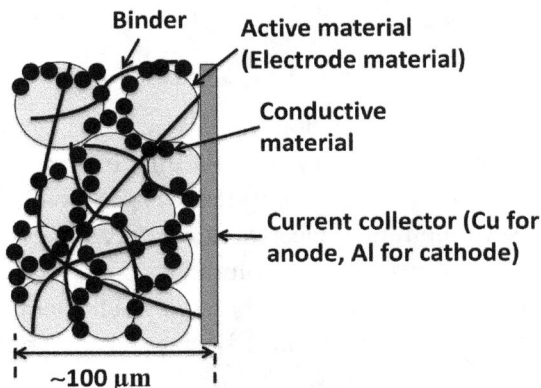

Fig. 2.2   Electrode of Li battery.

collectors are a thin foil of metal; Cu and Al foils are used as current collectors for the anode and cathode, respectively, because they are stable at the working potential of the electrodes. Al current collector can be also used for $Li_4Ti_5O_{12}$ anode. TOSHIBA commercialized the Li batteries using $Li_4Ti_5O_{12}$ as the anode in 2008 (SCiB) [7]. The working potential of $Li_4Ti_5O_{12}$ anode is 1.55 V vs. Li^+/Li [8], which is much higher than that of graphite, which is below 0.2 V vs. Li^+/Li [9]. Therefore, the Al current collector also can be used for the anode in this battery system, thereby reducing production cost. As for the cathode, $LiFePO_4$, $LiMn_2O_4$ and $LiCo_{1/3}Mn_{1/3}Ni_{1/3}O_2$ have also been commercialized.

To ensure high energy density, electrode materials should satisfy the following three basic requirements:

(1) High cell voltage, that is, electrode redox reaction proceeds at high (cathode) or low (anode) potential. This leads to high operation voltage of batteries.
(2) High specific charge and charge density, i.e. a high number of charge carriers per mass and volume unit of the material are available. This leads to high capacity of batteries.
(3) High reversibility, i.e. electrochemical reactions at both cathodes and anodes should maintain their reversibility for hundreds to

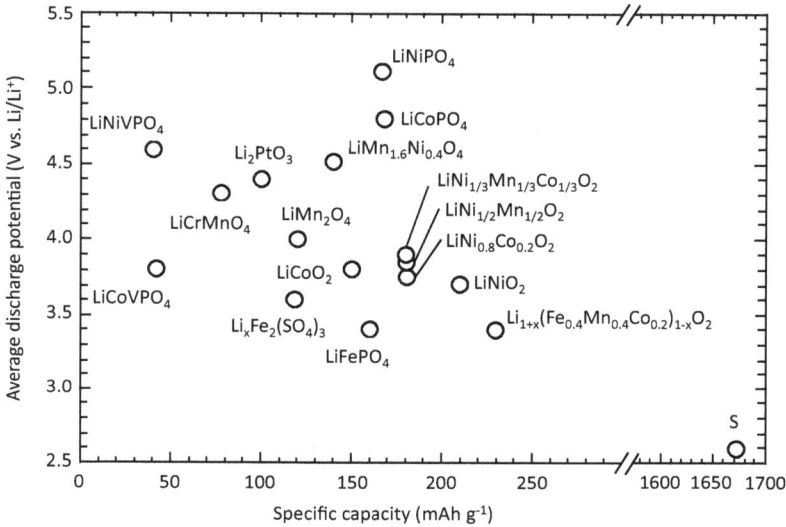

Fig. 2.3   Voltage and capacity of possible cathode materials.

thousands of charge–discharge cycles, resulting in stable charge and discharge capacities.

Most of the Li intercalation materials can be candidates for electrode materials in rechargeable Li batteries depending on their Li intercalation/deintercalation voltage. Figure 2.3 reveals the voltage and capacity of the possible cathode materials. Sulfur shows extremely high capacity compared with other cathode materials. Therefore, a Li battery using a sulfur cathode is usually categorized as a Li–S battery, to differentiate it from the Li battery. Intercalation electrodes are served as a host solid where the guest species (Li-ion) can be intercalated/deintercalated from/into the electrolyte. The group of intercalation materials is a special family of materials which have the capability to intercalate mobile guest species (atoms, molecules and ions) reversibly into a crystal host lattice. The host lattice possesses empty lattice sites with appropriate sizes for the guest species. In the intercalation materials, the volume of materials is not largely changed by the intercalation [10]. This leads to stable charge and discharge capacity for many cycles and also leads to the

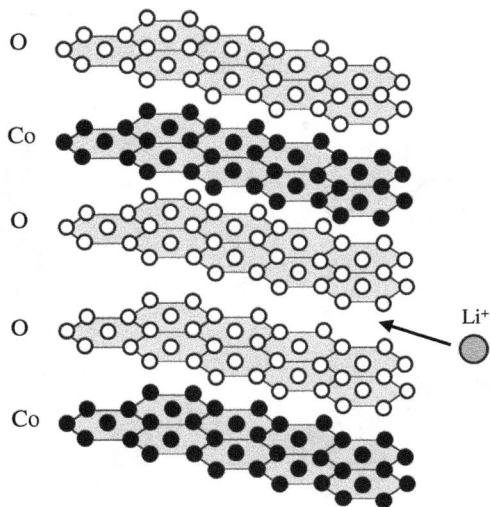

Fig. 2.4   Structure of LiCoO$_2$.

long life of the electrode. Although various host lattices such as metal dichalcogenides, metal oxyhalides, metal phosphorous trisulfides and so on have been reported thus far, only graphite and transition metal oxides have been exclusively used for the electrode materials of Li batteries.

Transition metal oxides have been used as cathode materials. LiCoO$_2$ cathode has been used since Li batteries was commercialized in 1991. The structure of LiCoO$_2$ is depicted in Fig. 2.4. LiCoO$_2$ is composed of Li and Co layers stacked alternatively in-between oxygen layers [11]. This structure is defined commonly by a hexagonal cell having rhombohedral symmetry with space group R-3m. The theoretical capacity of LiCoO$_2$ is 273 mA h g^{-1}. However, the actual specific capacity is limited to the range of 137–140 mA h g^{-1} [12]. This corresponds to $x \approx 0.5$ in Li$_x$CoO$_2$, because the structure of LiCoO$_2$ is collapsed in the range of $x < 0.5$. To improve usage of Li ion (x), LiCo$_{1/3}$Mn$_{1/3}$Ni$_{1/3}$O$_2$ was developed. By partial substitution of Co with Ni and Mn, the stability of the structure is enhanced, and thus capacity is increased. The transition metal oxides tend to release oxygen at about 200°C in the charged state ($x <$ ca. 0.5).

The released oxygen may react with flammable organic electrolytes causing serious safety issues such as explosion and fire hazards. Therefore, a safety device for inhibition of over-charge is installed in commercial Li batteries.

$LiMn_2O_4$ with spinel structure is also a popular cathode material. Mn is more abundant than Co and is non-toxic to human tissues. Thus, $LiMn_2O_4$ is considered to be suitable as an alternative to $LiCoO_2$. In principle, $Li_xMn_2O_4$ can intercalate/deintercalate Li ions in the range of $0 < x < 2$. In $1 < x < 2$; the material consists of two different phases, i.e. cubic in bulk and tetragonal at the surface. Intercalation of Li ion causes a reduction in the valence of Mn ion in $LiMn_2O_4$ ($Mn^{4+} \rightarrow Mn^{3+}$), leading to a pronounced cooperative Jahn–Teller effect, in which the cubic spinel crystal becomes distorted tetragonal with a $c/a \approx 1.16$ and the volume of the unit cell increases by 6.5% [13]. The transition metal oxides usually demonstrate poor rate capability due to low Li-ion diffusion constant and electronic conductivity. In order to improve the rate capability, i.e. the intercalation/deintercalation kinetics of the materials, a reduction of the particle size of $LiMn_2O_4$ and addition of excess amount of the carbon materials during electrode preparation have been attempted and recognized as good strategies for this purpose.

Li transition metal phosphates with olivine structure, such as $LiFePO_4$ and $LiMnPO_4$, have also been widely studied as cathode materials, one of which, $LiFePO_4$, has been already commercialized and used in electrical vehicles (BYD Auto Co. Ltd.). Phosphate-based cathode materials have a strong covalent bond between P and O atoms. Thus, the olivine materials are stable and do not release oxygen. This high structural stability of phosphate-based cathode materials largely improves the safety. However, a problem of the olivine materials associated with low Li ion and electronic conductivities is encountered. To resolve this issue, nanostructuring of powder size and coating of conductive layers on the surface of powder particles are introduced. Owing to the reduction of particles size to shorten Li-ion diffusion path and carbon coating to improve electronic conductivity, $LiFePO_4$ was successfully commercialized. The operation voltage of $LiFePO_4$ (3.4 V vs. Li^+/Li) is lower than that of $LiCoO_2$ and $LiMn_2O_4$ (Fig. 2.3), resulting in lower energy density when using $LiFePO_4$ as the cathode, even though the safety of the battery is enhanced. Therefore, other olivine materials such as $LiMnPO_4$ (operation voltage: 4.1 V), $LiCoPO_4$ (4.8 V) and $LiNiPO_4$ (5.1 V) have been intensively researched. In particular, the operation voltage of $LiMnPO_4$ is quite similar to the current $LiCoO_2$ and $LiMn_2O_4$ cathodes. This implies that other components of the current battery system can be used and the safety of battery can be improved without loss of energy density by using $LiMnPO_4$ cathode.

Anode materials can mainly be categorized into three groups depending on their reaction with Li. These are as follows:

(1) Intercalation/deintercalation materials such as carbon-based materials and $Li_4Ti_5O_{12}$.
(2) Alloy/de-alloy materials such as Si, Ge, Sn Al, etc.
(3) Conversion materials like transition metal oxides (Mn_xO_y, NiO, Fe_xO_y, CuO, Cu_2O, etc.), metal phosphides, metal sulfides and metal nitrides (M_xX_y, X = S, P, N). Some conversion materials, such as Sn, Si, Ge, Bi and so on, exhibit extremely high capacities due to the alloy formation, i.e. both the conversion reaction and the alloy reaction proceed simultaneously [14–19]. Such materials are normally categorized as conversion materials.

Carbon-based anode materials can intercalate Li ion and are further classified into two groups, soft carbon (graphitizable carbons) and hard carbon (non-graphitizable carbons). The soft carbon has a layered structure with the carbon atoms arranged in a honeycomb lattice in each layer. The carbon atoms bond with neighborhood carbon atoms by strong covalent bonds, while each layer is connected through weak van der Waals force (Fig. 2.5(a)). The Li ions are accommodated between the carbon layers. The soft carbon reveals stable reversible capacity (i.e. 350–370 mAh g^{-1}), long cycle life and good coulomb efficiency [20, 21] and has been used as the anode material for Li batteries for a long time. The theoretical capacity of the soft carbon (graphite) is almost achieved. To enhance energy density of the Li battery, novel anode materials have been studied.

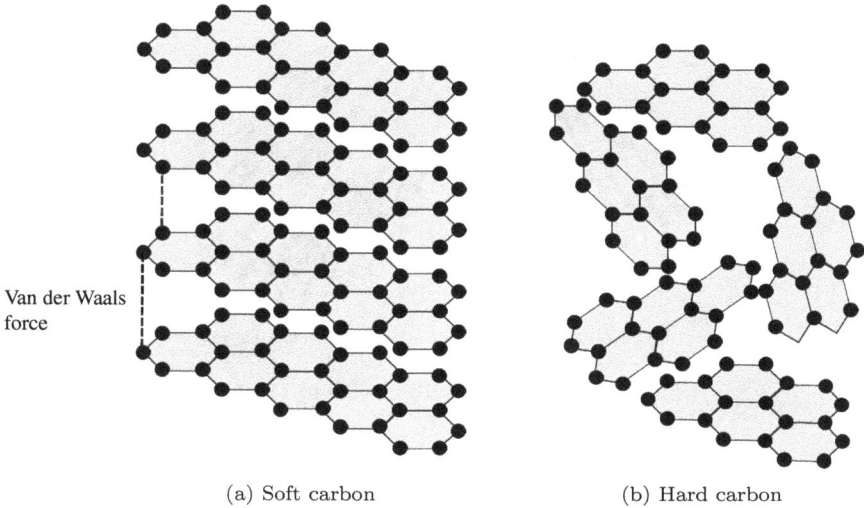

(a) Soft carbon   (b) Hard carbon

Fig. 2.5   Structure of soft carbon and hard carbon.

The hard carbons have higher reversible capacity (more than 500 mAh g^{-1}) than that of the soft carbons because the disordered orientations of the hard carbons can break the stoichiometry between Li and the carbon atoms (1:6 in the soft carbon). A shortcoming of the hard carbons is slow Li diffusion inside the hard carbons. To overcome this shortcoming, porous hard carbons have been studied [22].

Some other types of carbon material, such as carbon nanotubes (CNTs) and graphene, have also been expected to be used as anode materials. CNTs are allotropes of carbon with a cylindrical nanostructure and normally possess a length-to-diameter ratio of up to 132,000,000:1, which is significantly larger than that of other materials [23]. These cylindrical carbon molecules exhibit unusual properties which are valuable for nanotechnology, optics, electronics and other fields of materials science; especially, their superior electronic conductivity and good mechanical and thermal stabilities make them suitable for use as anode material [24]. There are two kinds of CNTs, that is, single (SWCNTs) and multiwall carbon nanotubes (MWCNTs), depending on the thickness and the number of coaxial layers (Fig. 2.6).

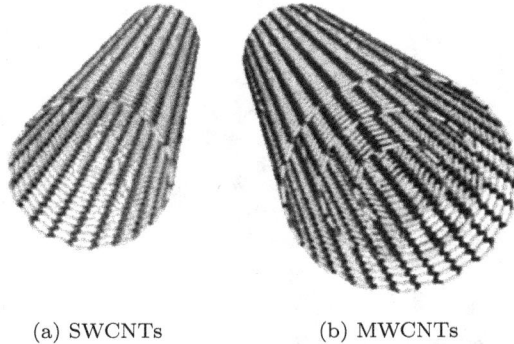

(a) SWCNTs   (b) MWCNTs

Fig. 2.6   Structure of SWCNT and MWCNT.

The theoretical capacity of SWCNT is estimated at 1116 mAh g^{-1} in LiC$_2$ stoichiometry, which is the highest among the carbon materials [25–27]. This high capacity is thought to be due to the intercalation of Li into stable sites on the surface of the nanotube which have a pseudo-graphitic layer. The high capacity of 1050 mAh g^{-1} was reported in purified SWCNTs prepared by laser vaporization [28]. However, obtaining a value closer to the theoretical one is still a major challenge in the development of CNT anode.

Graphene has also been considered to be a promising anode material for the next generation of Li batteries. Graphene is comprised in a honeycomb network of sp^2 carbon bonded into 2D sheets with nanometer thickness (single-atom thickness). In the other words, graphene is a single-layer graphite (Fig. 2.7). Graphene possesses high electrical conductivity and mechanical strength, high value of charge mobility and high surface area. These properties make graphene a good candidate for use as anode material [29, 30]. It is reported that high-quality graphite with a few graphene layers demonstrated large reversible capacity close to 1200 mAh g^{-1} [31]. Graphene/metal or metal oxide hybrid also has been studied. About 2–3 nm SnO$_2$/graphene hybrid showed high capacity and good stability (1220 mAh g^{-1} over 100 cycles) [32]. Similarly, Si/graphene and Fe$_3$O$_4$/graphene hybrids were also prepared and showed higher reversible capacity than that of the currently available graphite

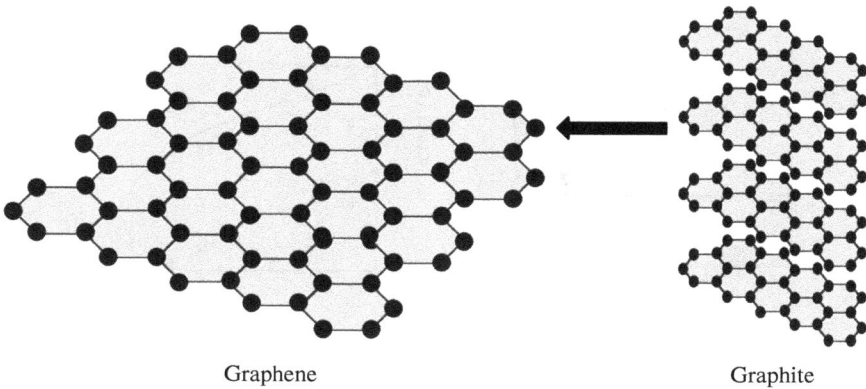

Graphene   Graphite

Fig. 2.7   Structure of graphene.

anode [33, 34]. The graphene can absorb the volume change of metal oxides due to its structural flexibility as well as provide a stable electrical conductive network. Also, the graphene acts as a Li intercalation material and contributes to the capacity. However, the high production cost of graphene and the formation of large amounts of solid electrolyte interface (SEI) layers currently restrict its application in Li batteries.

$Li_4Ti_5O_{12}$ anode with spinel structure has been considered the most appropriate titanium-based oxide for anode and was successfully commercialized by TOSHIBA Corporation in 2008. It should be mentioned that while Li insertion into/extraction from $Li_4Ti_5O_{12}$ occurs reversibly, the spinel symmetry and its structure remain almost unaltered (volume change is only 0.2–0.3%) [35–37]; therefore, this structure of $Li_4Ti_5O_{12}$ provides a remarkably long cycle life. Figure 2.8 depicts the typical charge and discharge curves of $Li_4Ti_5O_{12}$ and $LiCoO_2$ at 0.5 C. Compared with $LiCoO_2$, a long plateau can be seen at 1.55 V vs. Li/Li^+ in $Li_4Ti_5O_{12}$ and the Li insertion/extraction voltage is almost independent of capacity. The Li intercalation into $Li_4Ti_5O_{12}$ proceeds by a two-phase reaction in between $Li_4Ti_5O_{12}$ and $Li_7Ti_5O_{12}$. Both phases have the same crystallographical space group, Fd-3m [38]. Even upon intercalation of 3 Li^+ per formula unit, the original spinel structure is maintained

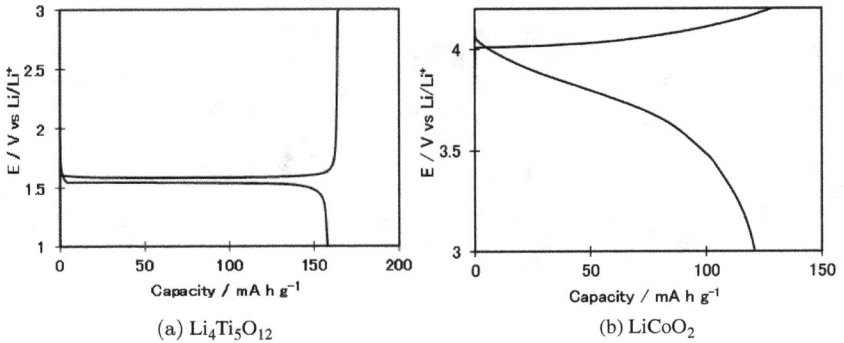

(a) $Li_4Ti_5O_{12}$ (b) $LiCoO_2$

Fig. 2.8   Charge and discharge curves of (a) $Li_4Ti_5O_{12}$ and (b) $LiCoO_2$ at $0.5°C$.

with only a small change in volume (0.2–0.3%) [39]. Therefore, $Li_4Ti_5O_{12}$ shows the long plateaus in charge and discharge curves and long cycle life. The high operation voltage of $Li_4Ti_5O_{12}$ is unfavorable as an anode because it reduces the operation voltage of Li batteries. However, the formation of a SEI and the deposition of Li dendrites, which are a typical issue in carbon-based anodes, can be avoided by using $Li_4Ti_5O_{12}$ anode [40]. The issue of low operation voltage of Li battery with $Li_4Ti_5O_{12}$ anode could be solved by using high-voltage cathode materials such as $LiNi_{0.5}Mn_{1.5}O_4$ (4.7 V) and $LiCoPO_4$ (4.8 V). However, in the present electrolyte system, decomposition of the electrolyte is inevitable over 4.5 V. Therefore, development of a novel electrolyte system is needed. The theoretical capacity of $Li_4Ti_5O_{12}$ is 175 mAh g^{-1}. However, low electronic conductivity ($\sim 10^{-13}$ S cm^{-1}) of $Li_4Ti_5O_{12}$ restricts the obtainment of the full capacity at high charge–discharge rates [41]. Carbon coating and nanostructuring of $Li_4Ti_5O_{12}$ have been researched to obtain the full capacity and improve the rate capability. For example, $Li_4Ti_5O_{12}$ nanowire was prepared directly on the Ti substrate. The nanowire demonstrated a good rate performance, i.e. 173 mAh g^{-1} at 0.2C and 121 mAh g^{-1} at $30°C$, and good cycle life. Another advantage of the nanowire on the Ti substrate is that Ti substrate can act as a current collector, and so binders are not needed [42].

Li-metal alloys, Li_xM_y, are of great interest as high-capacity anode because of their ability to store large amounts of Li [43, 44].

Table 2.2 Properties of alloy anodes.

Materials	Li	Graphite	Si	Sn	Sb	Al
Density (g cm^{-3})	0.53	2.25	2.33	7.29	6.7	2.7
Lithiated phase	Li	LiC$_6$	Li$_{4.4}$Si	Li$_{4.4}$Sn	Li$_3$Sb	LiAl
Theoretical capacity (mA h g^{-1})	3862	372	4200	994	660	993
Volume change	100	12	320	260	200	196
Potential vs. Li(V)	0	0.05	0.4	0.6	0.9	0.3

A comparison of some properties of alloy anodes with Li-metal and graphite anode is shown in Table 2.2. The theoretical capacities of the alloy anodes are normally 2–10 times higher than that of the present graphite anode. This high capacity is promising for next-generation Li batteries because it can reduce the volume of the anode and enhance the energy density. The second advantage of the alloy anodes is moderate operation voltage, which can avoid Li deposition under high current charge condition and then improve the safety of the battery. The main challenge of the alloy anodes for application to Li battery is their large volume change during alloying and de-alloying [45, 46]. This volume change provokes pulverization of the active alloy and poor cycle stability. Additionally, the first cycle irreversible capacity is too high for practical use. Many groups have extensively studied ways to solve these two issues, and significant progress has been achieved in the last two decades. In the studies, the downsizing from micro- to nanoscale particles, preparation of nanostructured porous alloy and fabrication of composite which is composed of Li active and inactive materials are recognized as the most promising ways. In the latter case, the Li inactive materials act as a buffer layer to absorb the stress due to volume change [47]. Many metals (Si, Sn, Sb, Al, Mg, Bi, In, Zn, Pb, Ag, Pt, Au, Cd, As, Ga and Ge) are active towards the alloy formation with Li [48]. However, only the first two elements have been extensively researched because they are cheap, abundant and environmentally friendly.

The last category of anode materials is the conversion material-pioneered by Poizot *et al.* [49]. The conversion reaction is generally described as follows:

$$M_aX_b + (b \cdot n)Li \leftrightarrow aM + bLi_nX$$

where M, X and n denote the transition metal, anion and formal oxidation state of X, respectively.

Normally, the transition metal compounds which do not contain any vacant sites are not suitable for use as electrode material because the compounds cannot intercalate Li ions. However, several transition metal oxides have been found to deliver larger stable charge and discharge capacity for Li ion than that of the currently available graphite anode; so, conversion materials have also been considered as one of the candidates for use as an anode. Since then, many other examples of the conversion reactions in sulfides, nitrides, fluorides and phosphides have been reported [50–55]. For example, the binary transition metal oxides with rock-salt structure (CoO, CuO, NiO, FeO) can react with Li reversibly. The reaction is as follows (Fig. 2.9):

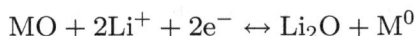

$$MO + 2Li^+ + 2e^- \leftrightarrow Li_2O + M^0$$

Their full reduction forms a composite material composing nanometer-range metallic particles (2–8 nm) dispersed in an amorphous Li_2O matrix. The nanometer character of the metal particles remains even after several reduction–oxidation cycles [56]. Li_2O was thought to be electrochemically inactive; therefore, the nanometer-sized metal particles would enhance the electrochemical activity of

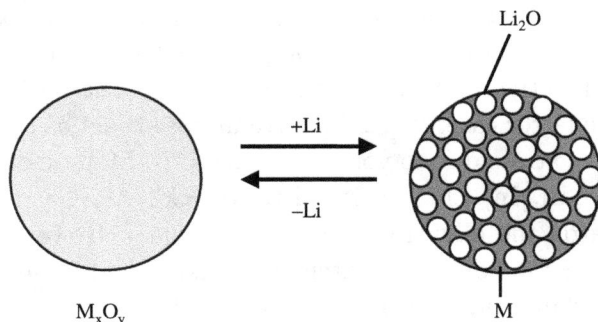

Fig. 2.9   Conversation reaction.

Li_2O toward formation and decomposition of Li_2O. It can be assumed that this phenomenon directly correlates with the valence of the M–O (M–X) bond. The redox potentials of conversion materials are tabulated in Table 2.3. The redox potential depends on the transition metal and anionic species, and so the redox potential can be tuned by choosing a suitable transition metal and anion species [57]. The conversion reaction proceeds at low potential as can be seen in Table 2.3. Only CuF_2 shows relatively moderate redox potential of 3.0 V vs. Li^+/Li. Thus, the conversion electrode has been considered as an alternative of commercial graphite anode. The conversion materials have remarkably large reversible capacity. For example, the theoretical capacities of Cr_2O_3 and CoS_2 are 1058 and 870 mA h g^{-1}, respectively. However, the conversion materials are still far from commercialization. Common issues faced with the conversion materials are the large volume change, like the alloy anodes mentioned above, during structural reorganization.

The large volume change causes the pulverization of the electrode materials and peeling off from the current collector. In order to deal with this issue, nanostructured conversion materials have been intensively studied [58–60].

The non-aqueous electrolyte has been used in the Li batteries because of the narrow electrochemical window of aqueous electrolytes. The electrochemical window of water ranges from 0 to 1.2 V vs. SHE. Although the voltage for Li insertion into $LiCoO_2$ is within the electrochemical window of water, the insertion voltage for graphite is beyond this window (Fig. 2.10). Therefore, water is decomposed in the Li battery if an aqueous electrolyte is used. The non-aqueous organic electrolytes possess wider electrochemical window and can thus be used for Li batteries. The non-aqueous electrolyte consists of a Li salt (e.g. $LiPF_6$) and organic solvents. The organic solvents are usually a mixture of cyclic solvent with high dielectric constant, such as ethylene carbonate (EC) and propylene carbonate (PC), and acyclic solvent with low viscosity, such as dimethyl carbonate (DMC), diethyl carbonate (DEC) and ethyl methyl carbonate (EMC) (Fig. 2.11).

Table 2.3   Redox potential of various conversion materials.

M	X = O		X = S		X = P		X = F		X = N	
	Phase	E_{conv}	Phase	E_{conv}	Phase	E_{conv}	Phase	E_{conv}	Phase	E_{conv}
M = Cr	Cr_2O_3	0.2	CrS	0.85			CrF_3	1.8	CrN	0.2
M = Mn	MnO_2	0.4	MnS	0.7	MnP_4	0.2				
	Mn_2O_3	0.3								
	MnO	0.2								
M = Fe	Fe_2O_3	0.8	FeS_2	1.5	FeP_2	0.3	FeF_3	2.0	Fe_3N	0.7
	Fe_3O_4	0.8	FeS	1.3	FeP	0.1				
	FeO	0.75								
M = Co	Co_3O_4	1.1	CoS_2	1.65–1.3	CoP_3	0.3	CoF_2	2.2	CoN	0.8
	CoO	0.8	$Co_{0.92}S$	1.4					Co_3N	1.0
			Co_9S_8	1.1						
M = Cu	CuO	1.4	CuS	2.0–1.7	CuP_2	0.7	CuF_2	3.0		
	Cu_2O	1.4	Cu_2S	1.7	Cu_3P	0.8				
M = Mo	MoO_3	0.45	MoS_2	0.6						
M = Ni	NiO	0.6	NiS_2	1.6	NiP_3	0.7–	NiF_2	1.9	Ni_3N	0.6

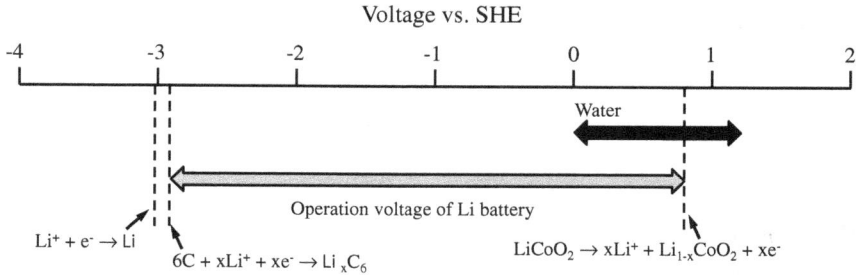

Fig. 2.10 Electrochemical window of water.

Cyclic solvent

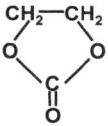

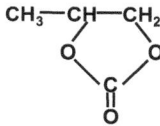

Acyclic solvent

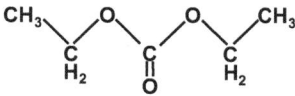

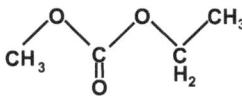

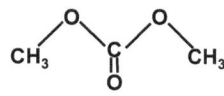

Fig. 2.11 Structures of organic solvents.

The conductivity of Li ion in the liquid electrolyte is expressed as

$$\sigma_{\mathrm{Li}} = zenu \tag{2.1}$$

where σ_{Li}, z, e, n, and u are the Li-ion conductivity, the valency of ion (Li ion $= 1+$), the elementary charge ($= 1.602 \times 10^{-19}$ C), and the density and mobility of the ion, respectively.

The mobility of the Li ion, u, follows the Nernst–Einstein equation

$$u = Dze/k_B T \tag{2.2}$$

wherein, D, k_B and T are the diffusion constant, Boltzmann constant ($= 1.381 \times 10^{-23}$ J K^{-1}) and absolute temperature, respectively.

Because the Li ion moves in the liquid electrolyte, the diffusion constant, D can be written by the Stokes–Einstein equation

$$D = k_B T/6\pi\eta r \tag{2.3}$$

Here, r and η are radius of the ion and viscosity of the liquid electrolyte, respectively.

From these equations, the Li-ion conductivity in the liquid electrolyte can be described as

$$\sigma = n(ze)^2/6\pi\eta r \tag{2.4}$$

Equation (2.4) clearly shows the Li-ion conductivity is associated with viscosity of the liquid electrolyte and the density of the Li ion (Li-ion concentration). Properties of representative organic solvents are summarized in Table 2.4.

EC and PC possess high dielectric constant, which promotes dissociation of the Li salt, resulting in increase of Li-ion concentration. However, high viscosity of these two solvents lowers Li-ion conductivity. To compensate the high viscosity of EC and PC, acyclic solvents with low viscosity such as DMC, DEC and EMC are usually added.

On the first charging of electrolyte containing EC, SEI is formed on the graphite anode. Figure 2.12 shows a typical discharge curve of graphite anode in the EC-containing electrolyte in the first cycle [61]. A small plateau can be observed at around 0.8 V. This voltage is much higher than that of intercalation voltage of Li ion into

Table 2.4   Properties of various organic solvents.

Solvent	Molecular weight	Dielectric constant	Viscosity (cp)	Melting point (°C)	Boiling point (°C)
EC	88.0	89.6	1.92	36.0	248.0
PC	102.0	64.4	2.53	−49.0	242.0
DMC	90.0	3.1	0.59	0.5	90.2
DEC	118.0	2.8	0.75	−43.0	126.0
EMC	104.0	3.0	0.65	−53.0	110.0

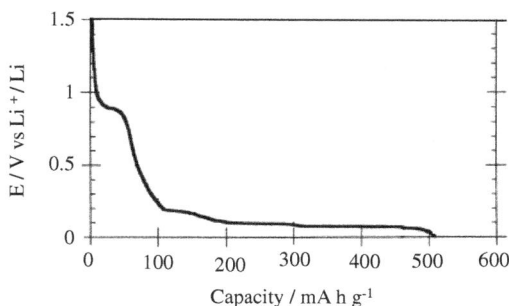

Fig. 2.12   Discharge curves of graphite anode in EC-containing electrolyte.

graphite and can be attributed to the formation of SEI. SEI is an *ad hoc* protective layer formed on the graphite anode surface by the decomposition of EC. The surface functional groups of graphite anode should influence the nature of the SEI layer. The type and distribution of surface functional group can enhance stability of the SEI layer [62] and also decomposition of the electrolyte, resulting in increasing the irreversible capacity [63, 64]. Therefore, surface functional groups on the graphite should be considered seriously.

By forming SEI, the graphite anode can intercalate and deintercalate Li ion reversibly and stably. Close value of discharge capacity to the theoretical capacity was already achieved at 3 C [65]. Here, 1 C denotes the current which can fully charge and discharge an active material (anode or cathode) for 1 h. For example, at 3 C, electrode is discharged and charged for 20 min (= 60 min/3) if the theoretical value is obtained. In fact, the electrode is usually charged and discharged within 20 min at 3C. Once the SEI layer has been deposited on the graphite surface, the layer hinders further decomposition of electrolyte. However, the layer allows Li ions to intercalate/deintercalate into/from the graphite anode.

Many attempts have been made to clarify the reaction pathways of the EC decomposition and formation of the SEI layer. However, the precise mechanisms of those reactions have not been made clear yet. It has been thought that the decomposition of EC that generates the common components of SEI layer proceeds through a two-electron reduction of a single molecule or two one-electron reductions of two

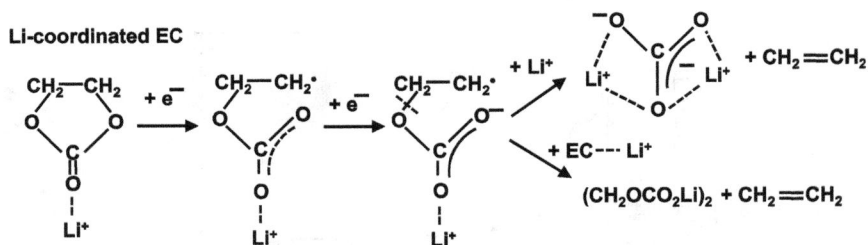

Fig. 2.13   Reduction pathway of EC.

molecules [66–68]. Also, computational studies have reported that the energy barrier for initial reduction of EC would be reduced to essentially eliminated level if the Li ion is coordinated with EC [69–72]. These results indicated that Li-coordinated molecules are preferentially reduced at the electrode surface. A probable pathway for EC decomposition was suggested by Aurbach *et al.* [73–75] (Fig. 2.13).

Li ions are solvated by one or more solvent molecules in the electrolyte. Coordination of solvent molecule(s) with a Li ion facilitates reduction of the solvent molecule(s) because this results in a negative free energy change for the first reduction reaction. To open the ring, there is an energy barrier. This energy barrier must be overcome to form the SEI layer. When the Li-coordinated cyclic carbonate adsorbs on the graphite surface, this additional coordination gives further stress to the ring structure of the carbonate. This facilitates the reduction of cyclic carbonates and the ring-opening step by lowering the energy barrier for ring opening [76, 77]. Furthermore, Li co-ordination would stabilize the radical anion formed by first one-electron reduction (Fig. 2.13(b)). The affinity for electron of anion is thought to be much lower than the cation and uncharged molecule; thus, it can be expected that the second reaction (Figs. 2.13(b) and 2.13(c)) does not occur immediately after the first reduction. Since these reactions proceed on the surface of carbon, the active surface of carbon would stabilize the radical anion. Once the second reduction occurs, the molecules react with either Li ion or Li-coordinated solvents. In both cases, ethene and Li carbonate or alkyl carbonate

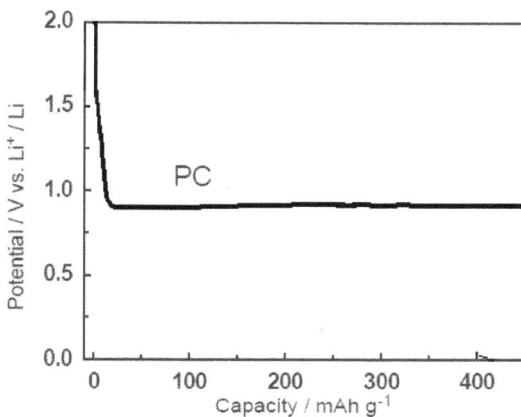

Fig. 2.14 Discharge curves of graphite anode in PC.

are formed. The Li carbonate and/or alkyl carbonate would be the main components of SEI.

EC is present in almost all commercial Li batteries because only EC can be decomposed in the first charging process and form the SEI layer. One of the problems of EC is its high melting temperature. This restricts the operation temperature of the batteries due to freezing of the electrolyte. Different from EC, PC has low melting temperature as well as high dielectric constant. Therefore, it has been expected to be suitable for use as a cyclic solvent. Figure 2.14 depicts a charge curve of the graphite anode obtained in PC containing $LiClO_4$ as the Li salt [78]. A large plateau at 0.8 V was observed in PC. This is due to exfoliation of the graphite. The Li intercalation into the graphite does not occur in the PC electrolyte because the SEI layer does not form on the graphite in the PC and the graphite anode is not protected. Thus, the PC cannot be used for Li batteries alone.

In order to develop stable SEI layer in the PC electrolyte, additives have also been researched. Most of the additives are functional derivatives of the cyclic carbonates [79–82]. Figure 2.15 shows structures of common additives.

In the various additives, vinylene carbonate (VC) is thought to be the most promising additive to form stable SEI layer [83–86]. The VC addition to electrolytes enhanced the long-term stability of

Vinylene carbonate (VC)	1,3-benzodioxol-2-one (BO)	Vinyl ethylene carbonate (VEC)	Vinyl ethylene Sulfite (VES)

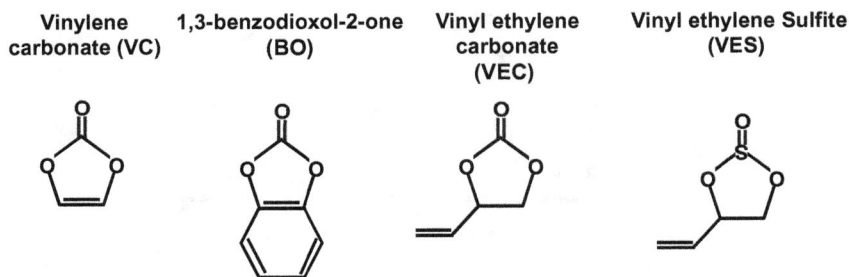

Fig. 2.15   Structures of additives.

the graphite anode. The VC is reduced and decomposed to species such as poly-Li-alkyl carbonate, $ROCO_2Li$ species and carbonates on the surface of the graphite anode. When graphite anode is polarized to low potential in the charging process, VC is reduced and forms more stable intermediates than the one formed by decomposition of EC [87]. The reduced VC is decomposed to radical anions, and these radical anions then undergo several termination reactions to produce alkyl dicarbonates, Li-carbides and so on. These reduction products may contain unsaturated double bonds. Therefore, they undergo further polymerization on the graphite surface and then build up a stable SEI film. Due to this promising nature, VC is expected to be an effective additive in non-EC based electrolytes.

It is critical to clarify the structure and properties of electrode–electrolyte interfaces as this largely affects the battery performance. The nature of the interface depends not only on the electrolyte solvent but also on the Li salts to a significant extent. The Li salt is also a key element to determine the performance of Li batteries. The ideal properties of salts for battery application are described as follows:

(1) The salt should have high solubility in the solvent.
(2) The counteranion should be stable toward oxidative and reductive decomposition on the both electrodes.
(3) It should be economic and easy to synthesize on a large scale.
(4) The counteranion should be non-toxic and thermally stable in battery working conditions.

(5) The counteranion should not react with the electrolyte solvent and other cell components.
(6) It should be non-toxic and non-flammable.
(7) It should be stable in ambient air for easy handling.

$LiPF_6$ has been widely used in commercial Li batteries. However, the low thermal stability of $LiPF_6$ has always evoked a safety concern, particularly in high-temperature operation. Additionally, $LiPF_6$ produces toxic HF when it reacts with water. Figure 2.16 depicts the structures of various Li salts, and the properties of Li salts are tabulated in Table 2.5 [88]. Among various Li salts, lithium bis(trifuluoromethanesulfonul) imide (LiTFSI) possesses a

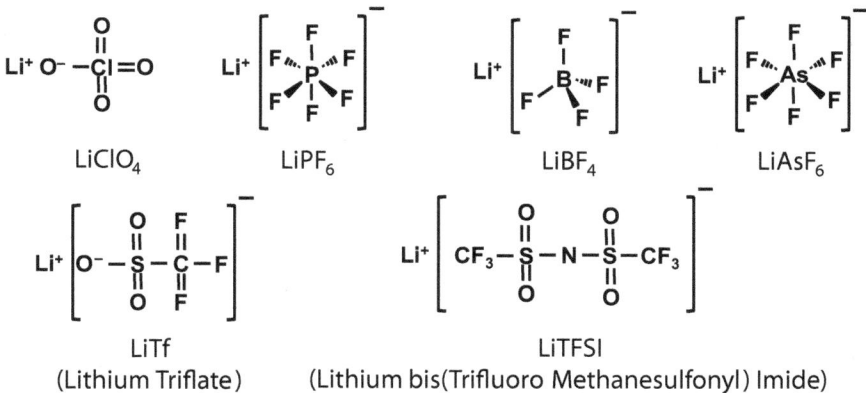

Fig. 2.16   Structures of various Li salts.

Table 2.5   Properties of various Li salts.

Property	From best → to worst					
Chemical inertness	LiTf	LiTFSI	$LiAsF_6$	$LiBF_4$	$LiPF_6$	
Solubility	LiTFSI	$LiPF_6$	$LiAsF_6$	$LiBF_4$	LiTf	
Ion pair dissociation	LiTFSI	$LiAsF_6$	$LiPF_6$	$LiClO_4$	$LiBF_4$	LiTf
Thermal stability	LiTFSI	LiTf	$LiAsF_6$	$LiBF_4$	$LiPF_6$	
Ion mobility	$LiBF_4$	$LiClO_4$	$LiPF_6$	$LiAsF_6$	LiTf	LiTFSI
SEI formation	$LiPF_6$	$LiAsF_6$	LiTFSI	$LiBF_4$		

Note: LiTf = lithium triflate, LiTFSI = lithium bis(trifluoro methanesulfonyl) imide.

good thermal stability and chemical stability. Therefore, much attention has been paid to this salt as a novel Li salt.

The development of liquid electrolytes have contributed to improving the performance of Li batteries together with the development of the electrode. However, the liquid electrolytes still have some issues such as flammability and narrow electrochemical window. The liquid electrolytes decompose over 4.5 V vs. Li^+/Li [89]. Therefore, high-voltage cathode materials such as $LiNi_{0.5}Mn_{1.5}O_4$ (4.7 V)[90] and $LiCoPO_4$ (4.8 V) [91] cannot be used with the current liquid electrolytes. Therefore, it is difficult to even test electrochemical properties of these high-voltage cathode materials. This narrow electrochemical window of the liquid electrolyte restricts the increase in energy density of the Li batteries.

Sulfone solvents are expected to be used as a high-voltage electrolyte for Li batteries. Sulfone-based electrolytes exhibited high stability of anodic potential above 6.0 V vs. Li^+/Li. In 1998, Angell *et al.* reported the possibility of using ethyl methyl sulfone (EMS) as a high-voltage solvent [92]. They reported EMS was stable above 5.5 V and the oxidative current, which is caused by oxidative cleavage of EMS, was not affected by the Li salt. However, most sulfones are solid at room temperature. Therefore, co-solvents with low melting points are needed to enable the usage of sulfone-based electrolytes. Angell further reported that by addition of DMC to EMS, a high oxidation limit of >5.9 V was obtained in EMS-DMC (1:1 by wt.) with 1 M $LiPF_6$. The initial coulomb efficiency reached 89.1% for $LiNi_{0.5}Mn_{1.5}O_4$ electrode, and the capacity retention was 97% over 100 cycles [94–96]. Another issue in the use of the sulfone-based electrolyte is exfoliation of graphite. Therefore, an additive for SEI formation is necessary when a graphite anode is used in the sulfone-based electrolyte. Lewandowski *et al.* added VC as the SEI formation additive to 1M $LiPF_6$/tetramethylene sulfone (TMS) [97]. The graphite anode delivered a capacity of 350 mA h g^{-1}. Moreover, $LiFePO_4$–1M $LiPF_6$/TMS+VC system demonstrated a flash point of about 150°C, which was much higher than that of the classical $LiFePO_4$-1M $LiPF_6$/EC-DMC system ($T_f \approx 37°C$). Also, usage of SEI-forming Li salts was studied [98–100]. Mao *et al.* found that lithium bis(oxalate)borate (LiBOB) could improve the capacity retention of a mesophase carbon microbeads anode. Moreover, $LiFePO_4$/Li cell with LiBOB-sulfolane/DMS electrolyte exhibited excellent cycling ability and good thermal stability at 60°C. Usage of a high-voltage anode such as $Li_4Ti_5O_{12}$ is also considered to be a possible way of avoiding the exfoliation of graphite in the sulfone-based electrolytes. It was reported that when $Li_4Ti_5O_{12}$ anode was used as anode in 1M $LiPF_6$/EMS-DMC electrolyte, the capacity retention was 100%; also, high coulombic efficiency was obtained [96].

Furthermore, the flammability of the liquid electrolyte is a serious concern in developing large-scale Li batteries for EV, HV and stationary energy storage. The large format batteries contain more liquid electrolyte, and so the safety issue becomes more serious. The flammability of the liquid electrolyte originates due to the presence of the organic solvents. Therefore, ionic liquids are being considered to replace the organic solvents. The ionic liquids are salts that are molten at low temperature (around room temperature) [101, 102]. It is well known that the ionic liquids show unusual properties such as negligible vapor pressure, high temperature stability and wide electrochemical window [103]. With ionic liquids as solvents, the safety characteristic of Li batteries could be greatly improved. Ionic liquids are composed of cations of large size such as pyridinium, imidazolium, quaternary ammonium, piperidium, pyrrolidinium, phosphonium, etc. and anions like $CF_3SO_3^-$, BF_4^-, PF_6^-, TFSI$^-$, etc. [104]. Common ion families in the ionic liquids are shown in Fig. 2.17 [105]. Among these cations, pyridinium and imidazolium cations are not stable with Li-metal. Therefore, the ionic liquids containing these cations are restricted in their use as solvents in Li batteries. The melting points of quaternary ammonium-, piperidinium-, pyrrolidinium- and phosphonium-based ionic liquids are lowered to room temperature only when they are combined with anions with large ionic radius like TFSI [106–116]. Sakaebe *et al.* studied the Li storage ability of TFSI anion based-ionic liquids using a Li/$LiCoO_2$ cell. The cathodic stability was found to

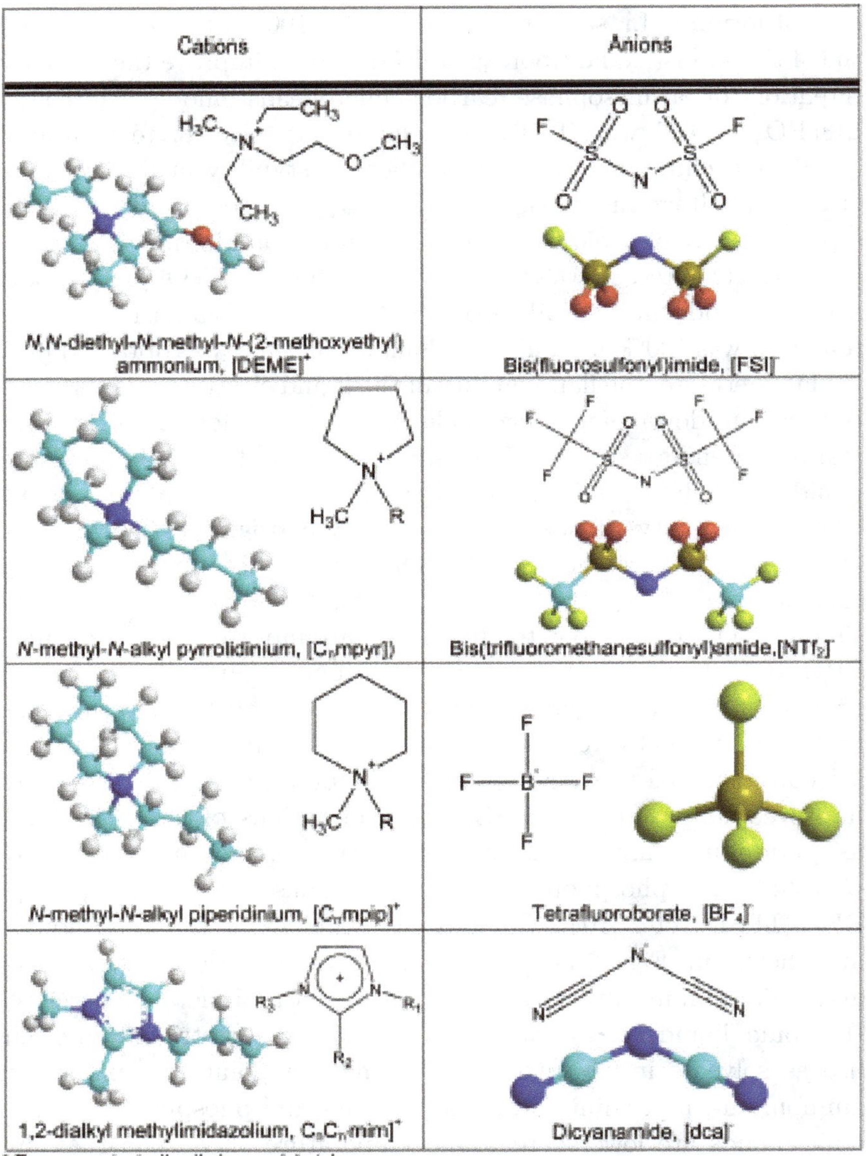

Cations	Anions
N,N-diethyl-*N*-methyl-*N*-(2-methoxyethyl) ammonium, [DEME]⁺	Bis(fluorosulfonyl)imide, [FSI]⁻
N-methyl-*N*-alkyl pyrrolidinium, [CₙMpyr])	Bis(trifluoromethanesulfonyl)amide,[NTf₂]⁻
N-methyl-*N*-alkyl piperidinium, [Cₙmpip]⁺	Tetrafluoroborate, [BF₄]⁻
1,2-dialkyl methylimidazolium, C₂Cₙmim]⁺	Dicyanamide, [dca]⁻

Fig. 2.17   Structures of ions used in ionic liquids.

Source: Reproduced from Ref. [105] with permission from Royal Society of Chemistry.

be EMI (1-ethyl-3-methyl imidazolium)TFSI < TMPA (trimethyl-propylammonium)TFSI < P13(N-methyl-N-propylpyridinium)TFSI ≈ PP13(N-methyl-N-propylpiperidinium)TFSI. The PP13TFSI was the most promising ionic liquid among the entire lot. The melting point and conductivity were 8.7°C and 1.51×10^{-3} S cm^{-1}, respectively. The Li/LiCoO$_2$ cell with PP13TFSI exhibited a capacity of 120 mA h g^{-1}, and the Coulomb efficiency during entire cycle was more than 97%. About 85% of the discharge capacity was retained after 30 cycles at 0.5C rate, whereas the same cell with TMPATFSI showed rapid capacity decay. When compared with conventional carbonate-based organic solvents, the cost of the ionic liquids is prohibitive [105]. Therefore, ionic liquids containing dicyanamide anion have been studied [117]. By avoiding the use of fluorinated anions, the cost of ionic liquids can be reduced. Figure 2.18 depicts the charge and discharge curves of Li/LiFePO$_4$ cell using pyrrolidinium dicyanamide ionic liquid at 1st, 50th and 100th cycle. The cell showed

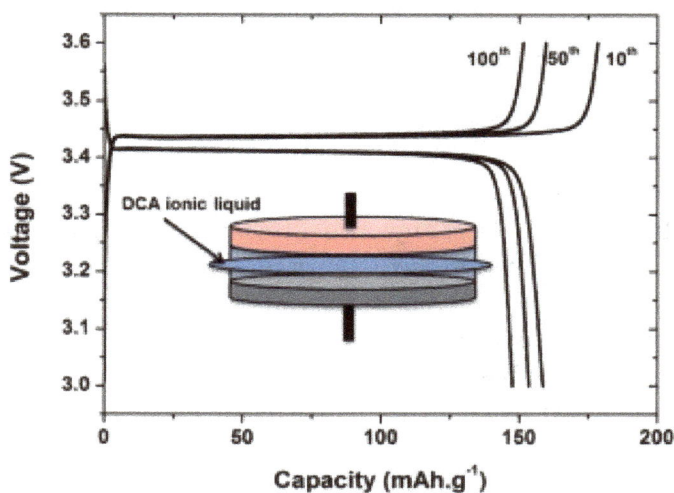

Fig. 2.18 Charge and discharge curves of 10th, 50th and 100th cycle of Li/LiFePO$_4$ cell using pyrrolidinium dicyanamide ionic liquid.

Source: Reproduced from Ref. [105] with permission from Royal Society of Chemistry.

a large plateau in both charge and discharge processes until the 100th cycle, indicating the high stability of the cell.

However, the graphite anode cannot work properly in many types of ionic liquids due to insufficient formation of SEI. Therefore, additives for SEI formation, such as VC, are needed. It was found that lithium difluoro(oxalate)borate (LiODFB) could form a stable SEI layer on the surface of a carbonaceous anode [107]. The SEI layer protects the carbonaceous anode from decomposition. The Li/MCMB cell exhibited a 1st discharge capacity of 369.5 $mA\,h\,g^{-1}$ in the LiODFB-PP_{14}TFSI/TMS cell. The cell still maintained a high discharge capacity of 338.6 mA h g^{-1} after 50 cycles. Sun *et al.* added 10 wt.% of VC to ionic liquids based on methhylpropylpyrrolidinium (MPPY) and methylpropylpiperridinium (MPPI) cations and TFSI anion [118]. Due to the addition of VC, a stable SEI layer was formed. The capacities were 325 and 310 mA h g^{-1}at 50°C for the cell using 0.5 M LiTFSI/MPPY-TFSI+10 wt.% VC and 0.5M LiTFSI/MPPI-RFSI+10 wt.% VC, respectively. The coulombic efficiency in both cells was above 96%, except for the first two cycles. It was also reported that by adding ether group, the graphite anode could work well in N,N-diethyl-N-methyl-N-(2-methoxyethyl) ammonium TFSI (DEME-TFSI) with LiTFSi salt electrolyte [108]. The graphite electrode demonstrated an initial capacity of 318 mA h g^{-1}. The discharge capacity was maintained at 320 mA h g^{-1}, and the coulombic efficiency was stable at 100%. Recently, stable charge and discharge behavior of graphite anode in the 1-ethyl-3-methylimidazolium bis(fluorosulfonyl)imide (EMImFSI)/LiTFSI electrolyte was reported. Graphite anode has a high-reversible capacity of 365 mA h g^{-1} [115]. As mentioned above, many ionic-based electrolytes have been synthesized and tested as electrolyte base for the Li batteries. However, there are still some issues to be overcome before it can be used in practice, such as lowering the high production cost, lowering the melting point, improving long-term stability and so on. The development of ionic liquid-based electrolytes is well reviewed in Refs. [119–122].

The electrolytes for next-generation Li batteries should possess high conductivity, wide electrochemical window to provide high energy and power densities, and also low (no) flammability for safety concerns. The solid ceramic electrolytes which are non-flammable and normally have wide electrochemical window ($\sim$6 V vs. Li^+/Li), meet these demands well. Therefore, they have been extensively researched to fabricate all-solid-state Li batteries, especially in the past two decades. Many solid electrolytes such as sulfides and oxides have been found so far, and their properties have also been studied. In the following chapters, the history, theory and properties of solid electrolytes are explained. Then, the application of the solid electrolytes to all-solid-state battery is described in the last chapter.

References

[1] M. Kotobuki, Y. Suzuki, K. Kanamura, Y. Sato, K. Yamamoto, T. Yoshida: A novel structure of ceramics electrolyte for future lithium battery, *Journal of Power Sources* 196 (2011) 9815–9819.

[2] C. J. Rydh, B. A. Sandén: Energy analysis of batteries in photovoltaic systems. Part I: Performance and energy requirements, *Energ. Convers. Manage.* 46 (2005) 1957–1979.

[3] J. Wang, Y. Li, X. Sun: Challenges and opportunities of nanostructured materials for aprotic rechargeable lithium-air batteries, *Nano Energy* 2 (2013) 443–467.

[4] K. Kanamura, M. Kotobuki: *New Material for Next Generation Rechargeable Batteries for Future Society* (CMC Publishing Co., Ltd., Tokyo, 2009).

[5] N. Rey-Raap, M.-L. C. Piedboeuf, A. Arenillas, J. A. Menendez, A. F. Leonard, N. Job: Aqueous and organic inks of carbon xerogels as models for studying the role of porosity in lithium-ion battery electrodes, *Mater. Design* 109 (2016) 282–288.

[6] X. Lu, G. Wu, Q. Xing, H. Qin, X. Lu, F. Luo: Fabrication of binder-free graphene-SnO_2 electrodes by laser introduced conversion of precursors for lithium secondary batteries, *Appl. Surface Sci.* 406 (2017) 265–273.

[7] Toshiba web-site: http://www.scib.jp/en/about/index.htm

[8] Y. Zhao, G. Liu, L. Liu, Z. Jiang: High-performance thin-film $Li_4Ti_5O_{12}$ electrodes fabricated by using ink-jet printing technique and their electrochemical properties, *J. Solid State Electrochem* 13 (2009) 70–711.

[9] J.-H. Lee, S. Lee, U. Paik, Y.-M. Choi: Aqueous processing of natural graphite particulates for lithium-ion battery anodes and their electrochemical performance, *J. Power Sources* 14 (2005) 249–255.

[10] D. O'Hare: Inorganic Materials, Wiley, New York (1991).

[11] Y. Shao-Horn, S. Levasseur, F. Weil, C. Dekmass: Probing lithium and vacancy ordering in O_3 layerer Li_xCoO_2 ($x \approx 0.5$), *J. electrochem. Soc.* 150(3) (2003) A366–A373.

[12] T. Ohzuku, A. Ueda: Why transition metal (di) oxides are the most attractive materials for batteries, *Solid State Ionics* 69 (1994) 201–211.

[13] M. M. Thackeray: Structural consideration of layered and spinel lithiated oxides for lithium ion batteries, *J. Electrochem. Soc.* 142 (1995) 2558–25663.

[14] P. Fiordiponti, G. Pistoia, C. Temperoni: Behavior of Bi_2O_3 as a cathode for lithium cells, *J. electrochem. Soc.* 125 (1978) 14–17.

[15] A. Huggins: Alloy negative electrodes for lithium batteries formed in-situ from oxides, *Ionics* 3 (1997) 245–255.

[16] Y. Idota, T. Kubota, A. Matsufuji, Y. Maekawa, T. Miyasita: Tin-based amorphous oxide: A high-capacity-ion-storage material, *Science* 276 (1997) 1395–1397.

[17] H. Li, X. J. Huang, L. Q. Chen: Anodes based on oxide materials for lithium rechargeable batteries, *Solid State Ionics* 123 (1999) 189–197.

[18] M. Martos, J. Morales, L. Snachez: Lead-based systems as suitable anode materials for Li-ion batteries, *Electrochim. Acta* 48 (2003) 615–621.

[19] B. K. Guo, J. Shu, Z. X. Wang, H. Yang, L. H. Shi, Y. N. Liu, L. Q. Chen: Electrochemical reduction of nano-SiO_2 in hard carbon as anode material for lithium ion batteries, *Electrochem. Commun.* 10 (2008) 1876–1878.

[20] C. C. Li, Y. W. Wang: Importance of binder compositions to the dispersion and electrochemical properties of water-based $LiCoO_2$ cathodes, *J. Power Sources* 227 (2013) 204–210.

[21] S. Boyanov, K. Annou, C. Villevieille, M. Pelosi, D. Zitoun, L. Monconduit: Nanostructured transition metal phosphide as negative electrode for lithium-ion batteries, *Ionics* 14 (2008) 183–190.

[22] J. Yang, X. Zhou, J. Li, Y. Zou, J. Tang: Study of nano-porous hard carbons as anode materials for lithium ion batteries, *Mater. Chem. Phys.* 135 (2012) 445–450.

[23] X. Wang, Q. Li, J. Xie, Z. Jin, J. Wang, L. Jinyong, J. Yang, F. Kaili, F. Shoushan: Fabrication of ultralong and electrically uniform single-walled carbon nanotubes on clean substrates, *Nano Lett.* 9 (2009) 3137–3141.

[24] Y. Yu, C. Cui, W. Qian, Q. Xie, C. Zheng, C. Kong, F. Wei: Carbon nanotube production and application in energy storage, *Asia-pacific J. Chem. Eng.* 8 (2013) 234–245.

[25] V. Meunier, J. Kephart, C. Roland, J. Bernholc: *Ab initio* investigations of lithium diffusion in carbon nanotube systems, *Phys. Rev. Lett.* 88 (2002) 075506.

[26] K. NIshidate, M. Hasegawa: Energetics of lithium ion adsorption on defecvive carbon nanotubes, *Phys. Rev. B* 71 (2005) 245418.

[27] J. Zhao, A. Buldum, J. Han, J. P. Lu: First-principles study of Li-intercalated carbon nanotube ropes, *Phys. Rev. Lett.* 85 (2000) 1706–1709.

[28] R. A. DiLeo, A. Castiglia, M. J. Ganter, R. E. Rogers, C. D. Cress, R. P. Raffaelle, B. J. Landi: Enhanced capacity and rate capability of carbon nanotube based anodes with Titanium contacts for lithium ion batteries, *ACS Nano* 4 (2010) 6121–6131.

[29] J. Hou, Y. Shao, M. W. Ellis, R. B. Moore, B. Yi: Graphene-based electrochemical energy conversion and storage: Fuel cells, supercapacitors and lithium ion batteries, *Phys. Chem. Chem. Phys.* 13 (2011) 15384–15402.

[30] X. Huang, X. Qi, F. Boey, H. Zhang: Graphene-based composites, *Chem. Soc. Rev.* 41 (2012) 666–686.

[31] P. Lian, X. Zhu, S. Liang, Z. Li, W. Yang, H. Wang: Large reversible capacity of high quality graphene sheets as an anode material for lithium-ion batteries, *Electrochim. Acta.* 55 (2010) 3909–3914.

[32] B. P. Vinayan, S. Ramaprabhu: Facile synthesis of SnO_2 nanoparticles dispersed nitrogen doped graphene anode material for ultrahigh capacity lithium ion battery applications, *J. Mater. Chem. A.* 1 (2013) 3865–3871.

[33] B. Wang, X. Li, X. Zhang, B. Luo, M. Jin, M. Liang, S. A. Dayeh, S. T. Picraux, L. Zhi: Adaptable silicon-carbon nanocables sandwiched between reduced graphene oxide sheets as lithium ion battery anodes, *ACS Nano* 7 (2013) 1437–1445.

[34] A. Hu, X. Chen, Y. Tang, Q. Tang, L. Yang, S. Zhang: Self-assembly of Fe_3O_4 nanorods on graphene for lithium ion batteries with high rate capacity and cycle stability, *Electrochem. Commun.* 28 (2013) 139–142.

[35] X. Li, C. Wang: Engineering nanostructured anodes vis electrostatic spray deposition for high performance lithium ion battery application, *J. Mater. Chem. A.* 1 (2013) 165–182.

[36] S. K. Martha, O. Haik, V. Borgel, E. Zinigrad, I. Exnar, T. Drezen, J. H. Miners, D. Surbach, $Li_4Ti_5O_{12}/LiMnPO_4$ lithium-ion battery systems for load levelling application, *J. Electrochem. Soc.* 158 (2011) A790–A797.

[37] G. Xu, P. Han, S. Dong, H. Liu, G. Cui, L. Chen: $Li_4Ti_5O_{12}$-based energy conversion and storage systems: Status and prospects, *Coord. Chem. Rev.* 343 (2017) 139–184.

[38] M. G. Verde, L. Baggetto, N. Balike, G. M. Veith, J. K. Seo, Z. Wang, Y. S. Meng: Elucidating the phase transformation of $Li_4Ti_5O_{12}$ lithiation at the nanoscale, *ACS Nano* 10 (2016) 4312–4321.

[39] T. Ohzuku, A. Ueda, N. Yamamoto: Zero-strain insertion material of $Li[Li_{1/3}Ti_{5/3}]O_4$ for rechargeable lithium cells, *J. Electrochem. Soc.* 142 (1995) 1431–1435.

[40] A. Mahmoud, J. M. Amarilla, K. Lasri, I. Saadoune: Influence of the synthesis method on the electrochemical properties of the $Li_4Ti_5O_{12}$ spinel in Li-half and Li-ion full-cells. A systematic comparison, *Electrochim. Acta.* 93 (2013) 163–172.

[41] G. N. Zhu, L. Chen, Y. G. Wang, C. X. Wang, R. C. Che, Y. Y. Xia: $Li_4Ti_5O_{12}$–$Li_2Ti_3O_7$ nanocomposite as an anode material for Li-ion batteries, *Adv. Funct. Mater.* 23 (2013) 640–647.

[42] L. Shen, E. Uchaker, X. Zhang, G. Cao: Hydrogenated $Li_4Ti_5O_{12}$ nanowire arrays for high rate lithium ion batteries, *Adv. Mater.* 24 (2012) 6502–6506.

[43] A. K. Shulka, T. P. Kumar: Materials for next-generation lithium batteries, *Curr. Sci.* 94 (2008) 314–331.

[44] M. Winter, J. O. Besenhard, M. E. Spahr, P. Novak: Insertion electrode materials for rechargeable lithium batteries, *Adv. Mater.* 10 (1998) 725–763.

[45] R. Benedek, M. M. Thackeray: Lithium reactions with intermetallic-compound electrodes, *J. Power Sources* 110 (2006) 406–411.

[46] U. Kasavajjula, C. Wang, A. J. Appleby: Nano- and bulk-silicon-based insertion anodes for lithium-ion secondary cells, *J. Power Sources* 163 (2007) 1003–1039.

[47] S. W. Woo, N. Okada, M. Kotobuki, K. Sasajima, H. Munakata, K. Kajihara, K. Kanamura: Highly patterned cylindrical Ni-Sn alloys with 3-dimensionally ordered macroporous structure as anodes for lithium batteries, *Electrochim. Acta* 55 (2010) 8030–8035.

[48] M. Winter, J. O. Besenhard: Electrochemical lithiation of tin and tin-based intermetallics and composites, *Electrochim. Acta* 45 (1999) 31–50.

[49] P. Poizot, S. Laruelle, S. Grugeon, L. Dupont, J. M. Trascon: Nano-sized transition-metal oxides as negative-electrode materials for lithium-ion batteries, *Nature* 407 (2000) 496–498.

[50] N. Pereira, L. Dupont, J. M. Trascon: Electrochemistry of Cu_3N with lithium: A complex system with parallel process, *J. Electrochem. Soc.* 150 (2003) A1273–A1280.

[51] H. Li, G. Ritchter, J. Maier: Reversible formation and decomposition of LiF clusters using transition metal fluorides as precursors and their application in rechargeable Li batteries, *Adv. Mater.* 15 (2003) 736–739.

[52] F. Badway, N. Pereira, F. Cosandey, G. G. Amatucci: Carbon-metal fluoride nanocomposites: Structure and electrochemistry of FeF_3:C, *J. Electrochem. Soc.* 150 (2003) A1209–A1218.

[53] B. Bervas, F. Badway, L. C. Klein, G. G. Amatucci: Bismuth fluoride nanocomposite as a positive electrode material for rechargeable lithium batteries, *Electrochem. Solid-State Lett.* 8 (2005) A179–A183.

[54] D. C. C. Silva, O. Crosnier, G. Ouvrard, J. Greedan, A. Safasefat, L. Nazar: Reversible lithium uptake by FeP_2, *Electrochem. Solid-State Lett.* 6 (2003) A162–A165.

[55] V. Pralong, D. C. S. Souza, K. T. Leung, L. Nazar: Reversible lithium uptake by CoP_3 at low potential: Role of the anion, *Electrochem. Commun.* 4 (2002) 516–520.

[56] S. Grugeon, S. Laruelle, L. Dupont, J. M. Tarascon: An update on the reactivity of nanoparticles Co-based compounds towards Li, *Solid State Sci.* 5 (2003) 895–904.

[57] J. B. Goodenough, Y. Kim: Challenges for rechargeable Li batteries, *Chem. Mater.* 22 (2010) 587–603.

[58] B. Wang, J. S. Chen, H. B. Wu, Z. Wang, X. W. Lou: Quasi emulsion-templated formation of α-Fe_2O_3 hollow spheres with enhanced lithium storage properties, *J. Am. Chem. Soc.* 133 (2011) 17146–17148.

[59] J. Liu, Y. Li, H. Fan, Z. Zhu, J. Jiang, R. Ding, Y. Hu, X. Huang: Iron oxide-based nanotube arrays derived from sacrificial template-accelerated hydrolysis: Large area design and reversible lithium storage, *Chem. Mater.* 22 (2009) 212–217.

[60] C. Wu, P. Yin, X. Zhu, C. Ouyang, Y. Xie: Synthesis of hematite (α-Fe_2O_3) nanorods: diameter-size and shape effects on their applications in magnetism, lithium ion battery and gas sensors, *J. Phys. Chem. B* 110 (2006) 17806–17812.

[61] S. Flandrois, B. Simon: Carbon materials for lithium-ion rechargeable batteries, *Carbon* 37 (1999) 165–180.

[62] P. Novak, D. Goers, M. E. Spahr: *Carbons for Electrochemical Energy Storage and Conversion Systems* (CRC Press, Florida, 2010).

[63] E. Buiel, J. R. Dahn: Li-insertion in hard carbon anode materials for Li-ion batteries, *Electrochimica Acta* 45 (1999) 121–130.

[64] C. E. Banks, T. J. Davies, G. G. Wildgoose, R. G. Compton: Electrocatalysis at graphite and carbon nanotube modified electrodes: Edge-plane sites and tube ends are the reactive sites, *Roy. Soc. Chem.* 0(7) (2005) 829–841.

[65] I. R. M. Kottegoda, Y. Kodama, H. Ikuta, Y. Uchimoto, M. Wakihara: Enhancement of rate capability in graphite anode by surface modification with zirconia, *Electrochem. Solid State Lett.* 5(12) (2002) A275–A278.

[66] D. Aubach, B. Markovskya, I. Weissmana, E. Levia, Y. Ein-Eli: On the correlation between surface chemistry and performance of graphite negative electrodes for Li ion batteries, *Electrochim. Acta* 45 (1999) 67–86.

[67] D. Ostrovskii, F. Ronci, B. Scrosati, P. Jacobsson: Reactivity of lithium battery electrode materials toward non-aqueous electrolytes: Spontaneous reactions at the electrode-electrolyte interface investigated by FT-IR, *J. Power Sources* 103 (2001) 10–17.

[68] M. Moshkovich, Y. Gofer, D. Aubach: Investigation of the elctrochemical windows of aprotic alkali metal (Li, Na, K) salt solutions, *J. Electrochem. Soc.* 148(4) (2001) E155–E167.

[69] T. Li, P. B. Balbuena: Theoretical studies of the reduction of ethylene carbonate, *Chem. Phys. Lett.* 317(3–5) (2000) 421–429.

[70] Y. Wang, S. Nakamura, M. Ue, P. B. Balbuena: Theoretical studies to understand surface chemistry on carbon anodes for lithium-ion batteries: Reduction mechanisms of ethylene carbonate, *J. Am. Chem. Soc.* 13(47) (2001) 11078–11718.

[71] H. Wang, M. Yoshio, T. Abe, Z. Ogumi: Characterization of carbon-coated natural graphite as a lithium-ion battery anode material, *J. Electrochem. Soc.* 149 (2002) A499–A503.

[72] Y. Wang, P. B. Balbuena: Theoretical insights into the reductive decompositions of propylene carbonate and vinylene carbonate: Density functional theory studies, *J. Phys. Chem. B* 106(17) (2002) 4486–4495.

[73] A. Naji, J. Ghanbaja, B. Humbert, P. Willmann, D. Billaud: Electroreduction of graphite in $LiClO_4$-ethylene carbonate electrolyte. Characterization of the passivating layer by transmission electron microscope and Fourier-transform infrared spectroscopy, *J. Power Sources* 63(1) (1996) 33–39.

[74] Y. Wang, P. B. Balbuena: *Lithium Ion Batteries: Solid-Electrolyte Interphase.* (Imperial College Press, London, 2004).

[75] R. Marom, O. Haik, D. Aurbach, I. C. Halalay: Revising LiClO4 as an electrolyte for rechargeable lithium-batteries, *J. Electrochem. Soc.* 157(8) (2010) A972–A983.

[76] Z. Wang, X. Huang, L. Chen: Performance improvement of surface-modified $LiCoO_2$ cathode materials: An infrared absorption and X-ray photoelectron spectroscopic investigation, *J. Electrochem. Soc.* 150(2) (2003) A199–A208.

[77] R. Marom, S. F. Amalraj, N. Leifer, D. Jacob, D. Aurbach: A review of advanced and practical lithium battery materials, *J. Mater. Chem.* 21(27) (2011) 9938–9954.

[78] M. Inaba, Z. Siroma, Y. Kawatate, A. Funabiki, Z. Ogumi: Electrochemical scanning tunneling microscopy analysis of the surface reactions on graphite basal plane in ethylene carbonate-based solvents and propylene carbonate, *J. Power Sources* 68 (1997) 221–226.

[79] X. G. Sun, S. Dai: Electrochemical investigations of ionic liquids with vinylene carbonate for applications in rechargeable lithium ion batteries, *Electrochim Acta* 55(15) (2010) 4618–4626.

[80] L. D. Xing, C. Y. Wang, M. Q. Xu, W. S. Li, Z. P. Cai: Theoretical study on reduction mechanism of 1,3-benzodioxol-2-one for the formation of solid electrolyte interface on anode of lithium ion battery, *J. Power Sources* 189(1) (2009) 689–692.

[81] W. Yao, Z. Zhang, J. Gao, J. Li, J. Xu, Z. Wang: Vinyl ethylene sulfite as a new additive in propylene carbonate-based electrolyte for lithium ion batteries, *Energy Environ. Sci.* 2(10) (2009) 1102–1108.

[82] Y. Ein-Eli, S. F. McDevitt, D. Aurbach, B. Markovsky, A. Schechter: Methyl propyl carbonate: A promising single solvent for lLi ion battery electrolytes, *J. Electrochem. Soc.* 144(7) (1997) L180–L184.

[83] Y. S. Yun, J. H. Kim, S. Y. Lee, E. G. Shim, D. W. Kim: Cycling performance and thermal stability of lithium polymer cells assembled with ionic liquid-containing gel polymer electrolytes, *J. Power Sources* 196(16) (2011) 6750–6755.

[84] S. Tsubouchi, Y. Domi, T. Doi, M. Ochida, H. Nakagawa, T. Yamanaka: Spectroscopic characterization of surface films formed on edge plane graphite in ethylene carbonate-based electrolytes containing film-forming additives, *J. Electrochem. Soc.* 159(11) (2012) A1786–1790.

[85] H. Kim, S. Grugeon, G. Gachot, M. Armand, L. Sannier, S. Laruelle: Ethylene bis-carbonates as telltales of SEI and electrolyte health, role of carbonate type and new additives, *Electrochim. Acta.* 136 (2014) 157–165.

[86] K. Hongyou, T. Hattori, Y. Nagai, T. Tanaka, H. Nii, K. Shoda, Dynamic in situ Fourier transform infrared measurements of chemical bonds of electrolyte solvents during the initial charging process in a Li ion battery, *J. Power Sources* 243 (2013) 72–77.

[87] Z. Wang, L. Zhou, X. W. Lou: Metal oxide hollow nanostructures for lithium-ion batteries, *Adv. Mater.* 24 (2012) 1903–1911.

[88] Y. Marcus: *Ion Solvation* (Wiley, New York, 1985).

[89] D. Aurbach, B. Markovsky, M. D. Levi, E. Levi, A. Schechter, M. Moshkovich, Y. Cohen: New insights into the intercalations between

electrode materials and electrolyte solutions for advanced nonaqueous batteries, *J. Power Sources* 81–82 (1999) 95–111.

[90] K. Hoshina, K. Yoshima, M. Kotobuki, K. Kanamura: Fabrication of $LiNi_{0.5}Mn_{1.5}O_4$ thin film cathode by PVP sol–gel process and its application of all-solid-state lithium ion batteries using $Li_{1+x}Al_xTi_{2-x}(PO_4)_3$ solid electrolyte, *Solid State Ionics* 209–210 (2012) 30–35.

[91] M. Kotobuki, Y. Mizuno, H. Munakata, K. Kanamura: Electrochemical properties of hydrothermally synthesized $LiCoPO_4$ as a high voltage cathode material for lithium secondary battery, *Phosphorus Res. Bull.* 24 (2010) 12–15.

[92] K. Xu, C. A. Angell: High anodic stability of a new electrolyte solvent: Unsymmetric noncyclic aliphatic sulfone, *J. Electrochem. Soc.* 145(4) (1998) L70–L72.

[93] X. G. Sun, C. A. Angell: New sulfone electrolytes: Part II. Cyclo alkyl group containing sulfones, *Solid State Ionics* 175 (2004) 257–260.

[94] X. G. Sun, C. A. Angell: New sulfone electrolytes for rechargeable lithium batteries: Part I. Oligoether-containing sulfones, *Electrochem, Comm.* 7 (2005) 261–266.

[95] L. G. Xue, S. Y. Lee, Z. F. Zhao, C. A. Angell: Sulfone-carbonate ternary electrolyte with further increased capacity retention and burn resistance for high voltage lithium ion batteries, *J. Power Sources* 295 (2015) 190–196.

[96] L. G. Xue, K. Ueno, S. Y. Lee, C. A. Angell: Enhanced performance of sulfone-based electrolytes at lithium ion battery electrodes, including the $LiNi_{0.5}Mn_{1.5}O_4$ high voltage cathode, *J. Power Sources* 262 (2014) 123–128.

[97] A. Lewandowski, B. Kurc, I. Stepniak, A. Swiderska-Mocek: Properties of Li-graphite and $LiFePO_4$ electrodes in Li PF_6-sulfolane electrolyte, *Electrochimica Acta* 56 (2011) 5972–5978.

[98] A. Hoffmann, T. Hanemann: Novel electrolyte mixtures based on dimethyl sulfone, ethylene carbonate and $LiPF_6$ for lithium-ion batteries, *J. Power Sources* 298 (2015) 322–330.

[99] A. Abouimrane, I. Belharouak, K. Amine: Sulfone-based electrolytes for high-voltage Li-ion batteries, *Electrochem. Comm.* 11 (2009) 1073–1076.

[100] L. P. Mao, B. C. Li, X. L. Cui, Y. Y. Zhao, X. L. Xu, X. M. Shi, S. Y. Li, F. Q. Li: Electrochemical performance of electrolytes based upon lithium bis(oxalate)borate and sulfolane/alkyl sulfite mixtures for high temperature lithium-ion batteries, *Electrochim. Acta* 79 (2012) 197–201.

[101] T. Welton: Room-temperature ionics liquids. Solvents for synthesis and catalysis, *Chem. Rev.* 99 (1999) 2071–2084.

[102] M. Galinski, A. Lewandowski, I. Stepniak: Ionics liquids as electrolytes, *Electrochim. Acta* 51 (2006) 5567–5580.

[103] M. Armand, F. Endres, D. R. MacFarlane, H. Ohno, B. Scrosati: Ionics-liquid materials for the electrochemical challenges of the future, *Nat. Mater.* 8 (2009) 621–629.

[104] P. Hapiot, C. Lagrost: Electrochemical reactivity in room-temperature ionic liquids, *Chem. Rev.* 108 (2008) 2238–2264.

[105] D. R. MacFarlane, N. Tachikawa, M. Forsyth, J. M. Pringle, P. C. Howlett, G. D. Elliott, J. H. Davis Jr., M. Watanabe, P. Simon, C. A. Angell: Energy applications of ionic liquids, *Energy Environ. Sci.* 7 (2014) 232–250.

[106] S. Pandian, S. G. Raju, K. S. Hariharan, S. M. Kolake, D. Park, M. Lee: Functionalized ionic liquids as electrolytes for lithium-ion batteries, *J. Power Sources* 286 (2015) 204–209.

[107] F. Wu, Q. Z. Zhu, R. J. Chen, N. Chen, Y. Chen, L. Li: Ionic liquid electrolytes with prospective lithium difluoro(oxalate)borate for high voltage lithium-ion batteries, *Nano Energy* 13 (2015) 546–553.

[108] J. Towada, T. Karouji, H. Sato, Y. Kadoma, K. Shimada, K. Ui: Charge-discharge characteristics of natural graphite electrode in N,N-diethyl-N-methyl-N-(2-methoxyethyl)ammonium bis(trifluoromethylsulfonyl) amide containing lithium ion for lithium-ion secondary batteries, *J. Power Sources* 275 (2015) 50–54.

[109] M. Yamagata, N. Nishigaki, S. Nishishita, Y. Matsui, T. Sugimoto, M. Kikuta, T. Higashizaki, M. Kono, M. Ishikawa: Charge-discharge behavior of graphite negative electrodes in bis(fuluorosulfonyl) imide-based ionic liquid and structural aspects of their electrode/electrolyte interfaces, *Electrochim. Acta* 110 (2013) 181–190.

[110] H. Sakaeba, H. Matsumoto, K. Tatsumi: Application of room temperature ionic liquids to Li batteries, *Electrochim. Acta* 53 (2007) 1048–1054.

[111] G. G. Eshetu, M. Armad, H. Ohno, B. Scrosati, S. Passerini: Ionic liquids as tailored media for the synthesis and processing of energy conversion materials, *Energy Environ. Sci.* 9 (2016) 49–61.

[112] M. V. Fedorov, A. A. Kornyshev: Ionics liquids at electrified interfaces, *Chem. Rev.* 114 (2014) 2978–3036.

[113] T. Nakazawa, A. Ikoma, R. Kido, K. Ueno, K. Dokko, M. Watanabe: Effects of compatibility of polymer binders with solvate ionic liquid electrolyte on discharge and charge reactions of lithium-sulfur batteries, *J. Power Sources* 307 (2016) 746–752.

[114] J. S. Moreno, Y. Deguchi, S. Panero, B. Scrosati, H. Ohno, E. Simonetti, G. B. Appetecchi: N-alkyl-N-ethylpyrrolidinium cation-based ionic liquid electrolytes for safer lithium battery systems, *Electrochim. Acta* 191 (2016) 624–630.

[115] Y. Matsui, M. Yamagata, S. Murakami, Y. Saito, T. Higashizaki, E. Ishiko, M. Kono, M. Ishikawa: Design of an electrolyte composition for stable and rapid charging-discharging of a graphite negative electrode in a bis(fluorosulfonyl)imide-based ionic liquid, *J. Power Sources* 279 (2015) 766–773.

[116] H. Sakaebe, H. Matsumoto: N-methyl-N-propylpiperidiniumbis(trifuluoromethanesulfonyl)imide (PP13-TFSI)-Novel electrolyte base for Li battery, *Electrochem. Commun.* 5 (2003) 594–598.

[117] H. Yoon, G. Lane, Y. Shekibi, P. Howlett, M. Forsyth, A. Best, D. MacFarlane: Lithium electrochemistry and cycling behavior of ionic liquids using cyano based anions, *Energy Environ. Sci.* 6 (2013) 979–986.

[118] X. G. Sun, S. Dai: Electrochemical investigations of ionic liquids with vinylene carbonate for applications in rechargeable lithium ion batteries, *Electrochim. Acta* 55 (2010) 4618–4626.

[119] A. Lewandowski, A. Swiderska-Mocek: Ionic liquids as electrolytes for Li-ion batteries-An overview of electrochemical studies, *J. Power Sources* 194 (2009) 601–609.

[120] I. Osada, H. de Vries, B. Scrosati, S. Passerini: Ionics-liquid-based polymer electrolytes for battery applications, *Angew. Chemie* 55 (2016) 500–513.

[121] M. Kar, T. J. Simons, M. Forsyth, D. R. MacFarlane: Ionic liquid electrolytes as a platform for rechargeable metal-air batteries: A perspective, *Phys. Chem. Chem. Phys.* 16 (2014) 18658–18674.

[122] A. Eftekhari, Y. Liu, P. Chen: Different roles of ionic liquids in lithium batteries, *J. Power Sources* 334 (2016) 221–239.

Chapter 3

History of Solid Electrolyte

Ion conduction in solids has been known for more than a century [1]. The first electrical conductive solid was discovered by Michael Faraday who, in 1883, reported that the electrical conductivity of Ag_2S was largely increased with increase in temperature [2]. Additionally, he discovered similar behavior in several other inorganic solids such as PbF_2, in 1838 [3]. This behavior was difficult to explain at that time because the behavior was opposite to that in metallic phase. Later, Hittorf investigated the conductivity of Ag_2S and Cu_2S and concluded that these materials had an electrolytic conduction mechanism [4]. Then, in 1884, Warburg proved that sodium ion could migrate through the solid, which overturned the conclusion of Arrhenius that neither pure salt nor pure water can be a conductor but that only salt dissolved in water.

One of the earliest applications employing solid electrolytes was the Nernst glower, which was an electronic light device containing oxide ion conductor, a mixture of ZrO_2 with rare-earth metal oxides as a light source [5]. The most favorable composition of the oxide ion conductor was 85% ZrO_2 and 15% Y_2O_3, which is well-known as Yttria-Stabilized Zirconia (YSZ). In 1960s, it was found that CeO_2–La_2O_3 solid solution showed higher ionic conductivity than YSZ and that the solid solution decreased overpotential when it was applied as a solid electrolyte in solid oxide fuel cells (SOFCs) [6]. Even though more than 100 years have passed since its discovery by Nernst, YSZ is still one of the most important oxygen electrolytes, and other alternative such as CeO_2 are still under investigation.

In parallel to developing the solid-oxygen ion conductors, remarkable high ionic conductivity of α-phase AgI was found in 1914 and only silver ion was identified as a mobile specie [7]. The α-phase AgI appears only above the phase transition at 419 K. In this phase, only anions (I^-) form the rigid lattice. On the other hand, cations (Ag^+) can migrate with low energy barriers, resulting in excellent ion conduction of the AgI. Since then, many Ag conductors with high conductivities at room temperature have been investigated [8–11]. Among them, the most impressive material is $RbAg_4I_5$. $RbAg_4I_5$ possesses excellent Ag ion conductivity of $0.21\,S\,cm^{-1}$, which is even $0.08\,S\,cm^{-1}$ higher than that in 1.0N NaCl aqueous solution at room temperature [12].

At the end of the 1960s, two novel inorganic solid electrolytes were found, and they received much attention due to their high ionic conductivity and potential applications in electrochemical devices such as batteries and capacitors. One was $RbAg_4I_5$ with high Ag ion conductivity, already mentioned above. Another was $Rb_4Cu_{16}I_7Cl_{13}$, which is a Cu ion conductor, and this is the solid electrolyte with the highest ionic conductivity to date [13]. In 1973, it was reported that a glass system was also a fast ion conductor with Ag^+ [14], which extended the development of solid electrolytes to the glass field as well. However, application of these Ag and Cu ion conductors in electrochemical devices were not successful despite their excellent high conductivity due to the very low decomposition potential (0.5–0.7 V), deliquescence in humid air and their costs. Moreover, a composition of excess Cu usually makes free electrons or holes, leading to the formation of mixed conductors.

Because of above difficulties to apply these solid electrolytes in electrochemical devices, much attention has gradually been shifted to alkali metal solid electrolytes. In 1967, β-Al_2O_3 was found as a Na ion conductor [15]; β-Al_2O_3 has a two-dimensionally layered structure, and Na ion migrates between the layers. This is the same as Li-ion conduction in graphite and $LiCoO_2$, which are used as anode and cathode in rechargeable Li batteries, respectively, in commercial

Li batteries. β-Al_2O_3 has been successfully commercialized in Na–S battery by NGK insulators, Ltd as mentioned in Chapter 1. The Na–S battery is used for storage of energy generated by solar power and wind power and as a source of emergency power supply. In 1976, Hong and Goodenough [16, 17] reported a sodium (Na) superionic conductor, named as NASICON, the structure of which consists of a rigid 3D framework interwoven with a three-dimensionally linked interstitial space, where Na ions can move with a low energy barrier. The 3D ionic pathways are preferred because they can theoretically avoid the conductivity degradation at boundaries of crystal grain in polycrystalline ceramics caused by anisotropic conduction.

Lithium is the lightest alkali metal with the smallest atomic number, indicating that an extremely high energy density can be obtained in Li batteries. This has stimulated us to develop solid electrolytes for Li battery. The first solid electrolyte, LiI, was reported in 1969 [18, 19]. Although the Li-ion conductivity of LiI was low (about $10^{-6} - 10^{-7}$ S cm^{-1}) [20], the conductivity was considerably enhanced by insulating Al_2O_3 addition [21]. The maximum conductivity, 4×10^{-5} S cm^{-1} at 298 K, was obtained in a composition of 35 mol% Al_2O_3 in LiI [22]. The conductivity enhancement was almost proportional to the total surface area of the dispersed Al_2O_3 particles. The "space charge layer" theory proposed by Wagner [23] and Maier [24] can explain this abnormal behavior. LiI has been used as solid electrolyte in commercialized cardiac pacemakers [25]. However, LiI is extremely hygroscopic and softens at about 100°C, limiting its wide application. Li_3N was another Li-ion conductor investigated intensively over the same period as LiI, because it exhibits much higher conductivities (over 10^{-3} S cm^{-1} at RT for its single crystals) [26]. However, Li_3N was shown to be unstable under potentials higher than 0.44 V [27] or in contact with atmosphere [28]. Additionally, the 2D conduction nature enhanced grain boundary resistance, resulting in changing the conductivity of polycrystalline Li_3N to orders of magnitude lower than that of its single crystalline form [29]. $LiBH_4$ was also proposed as a new solid electrolyte because of its high electrochemical stability, at least up to 5 V (vs. Li^+/Li) [30]. However, the high Li-ion conductive phase appears only above $120°C$, where the insulative low-temperature phase transforms into a conductive high-temperature phase [31]. Therefore, solid electrolytes with high electrochemical stability and high Li-ion conductivity at room temperature (RT) are required.

In 1977, NASICON compounds with a general formula $LiM^{I}M^{II}(PO_4)_3$ were developed as solid electrolytes with high Li-ion conductivity [32]. The 3D $[M^{I}M^{II}(PO_4)_3]$ skeleton of NASICON is composed of MO_6 octahedra linked to PO_4 tetrahedra. Li-ions are distributed in-between MO_6 and PO_4 and can move with low energy barrier. Abundant varieties of M^{I} and M^{II} are capable of forming the NASICON structure. Many NASICON compounds have been prepared and characterized so far. Another Li superionic conductor (LISICON) with 3D conduction pathway, $Li_{14}Zn(GeO_4)_4$, was developed by Hong in 1978 [33]. However, the pure solution phases for this kind of oxysalt can be formed only within limited ranges of composition and temperature, leading to difficulty in preparation of this kind of material [34]. Materials with perovskite structure, which originated from studies on dielectric and ferroelectric properties, were also found to have Li-ion conduction abilities. Inaguma *et al.* studied the Li-ion conductivity of $Li_{0.34}La_{0.51}TiO_{2.94}$ and reported a high bulk conductivity of $1 \times 10^{-3}\,S\,cm^{-1}$ [35]. This makes the perovskite-type oxide electrolytes increasingly interesting, thus leading to them being intensively investigated. A new type of Li-ion conductor which has a garnet structure, $Li_5La_3M_2O_{12}$ (M = Ta, Nb), was reported in 2003 [36]. To increase conductivity of the ion conductors with garnet structure, various ions have been doped and substituted at the La-site and M-site. The garnet compound in which the M site was replaced by Zr was found to have high Li-ion conductivity of $4 \times 10^{-4}\,S\,cm^{-1}$ [37]. Additionally, it showed excellent electrochemical stability. This type of solid electrolyte has been given much attention, and exhaustive research on these compounds is going on.

Sulfide electrolytes are attractive mainly due to their high Li-ion conductivity. Pure sulfide electrolyte has a similar structure as LISICON, so it is termed thio-LISICON [38]. $Li_{3.25}Ge_{0.25}P_{0.75}S_4$ exhibits the highest Li-ion conductivity among all thio-LISICON compounds [39]. The design concept can be extended to Li–Si–P–S compounds [40, 41]. This system is one of the most widely investigated ones in the solid electrolytes [42, 43]. Regarding other sulfide electrolytes, high Li-ion conductivity of argyrodite-type crystal, Li_6PS_5X (X = Cl, Br, I) was recently reported [44]. A remarkable innovation in the field of sulfide electrolytes appeared in 2011. A novel sulfide crystalline electrolyte with composition $Li_{10}GeP_2S_{12}$ showed a conductivity of 1.2×10^{-2} S cm^{-1} at room temperature; this was the highest conductivity among all the inorganic solid electrolytes thus far [45]. This high conductivity is even comparable to that of organic liquid electrolytes used in commercial Li batteries.

At the same time, amorphous/glass electrolytes have also been developed. Li_2O–SiO_2–B_2O_3 glass exhibited high Li-ion conductivity $> 10^{-4}$ S cm^{-1} at 350°C, as reported by Otto [46]. After that, various approaches such as increasing Li_2O concentration [47, 48] and adding Li salts [49–51] have been attempted to improve the Li-ion conductivity of glass electrolytes. By these approaches, the Li-ion conductivity at room temperature of higher than 10^{-6} S cm^{-1} has been achieved. However, this is still too low to apply for solid electrolytes. The most successful progress in amorphous/glass electrolyte is the development of $Li_xPO_yN_z$, known as LiPON. An amorphous LiPON with micron-level thickness was deposited by magnetron sputtering [52]. This small thickness of LiPON can compensate for the negative effect of the low conductivity of about 3.3×10^{-6} S cm^{-1} [53], and the thin-film LiPON was successfully used in a thin-film all-solid-state battery [54]. Although the LiPON-based thin-film all-solid-state battery was already commercialized by CYMBET Corporation [55], most other oxide glass electrolytes still have problems with the application of all-solid-state batteries due mainly to its low conductivity. However, sulfide glasses such as LiI–Li_2S–P_2S_5 [56] and LiI–Li_2S-B_2S_3 [57] were already found to have high conductivities of $\sim 10^{-3}$ S cm^{-1} in the early 1980s. After that, the research on sulfide glass shifted to LiI–Li_2S–SiS_2 glass [58] or Li_2S–P_2S_5 glass ceramics [59] because of the ease of preparation.

As mentioned above, promising Li-ion conductive ceramics are categorized into crystalline group, amorphous/glass group and glass ceramics group. Each group can be further categorized into oxide and sulfide groups. Detailed explanations of each group will be detailed in Chapters 5 and 6.

References

[1] P. Knauth, H. L. Tuller: Solid-state ionics: roots, status, and future prospects, *J. Am. Ceram. Soc.* 85(7) (2002) 1654–1680.

[2] M. Faraday: *On Conducting Power Generally; in Experimental Researches in Electricity*, Series IV (Royal Institution, London, 1833), pp. 119–126.

[3] M. Faraday: *Effect of Heat on Conduction; in Experimental Researches in Electricity*, Series XII (Royal Institution, London, 1839), pp. 436–450.

[4] W. Hittorf: *Ueber das elektrische Leitungsvermoegen des Schwefelsilbers und Halbschwefelkupfers*, *Ann. Phys. Chem.* 84 (1851) 1–28.

[5] W. Nernst: Material for Electric-Lamp Glowers, U.S. Pat. No. 685 730, 1901.

[6] T. Takahasi, K. Ito, M. Iwahara: The fuel cell with a new type solid electrolyte *Rev. Energie Primaire, Journées Int. d'Etude des Piles à Combustible, Bruxelles* 3 (1965) 42–48.

[7] C. Tubandt, E. Lorenz: *Molekularzustand und elektrisches eitvermoegenkristallisierter Salze*, *Z. Phys. Chem.* 87 (1914) 513–542.

[8] B. Reuter, K. Hardel: *Über die Hochtemperaturmodifikation von Silbersufidjodid*, *Naturwissenschaften* 48 (1961) 161.

[9] B. B. Owens, G. R. Argue: High-conductivity solid electrolytes: MAg_4I_5, *Science* 157 (1967) 308–310.

[10] T. Takahashi, S. Ikeda, O. Yamamoto: Solid-state ionics-solids with high ionic conductivity in the systems silver iodide-silver oxyacid salts, *J. Electrochem. Soc.* 119 (1972) 477–482.

[11] M. Tatsumisago, Y. Shinkuma, T. Minami: Stabilization of superionic α-AgI at room temperature in a glass matrix, *Nature* 354 (1991) 217–218.

[12] D. O. Raleigh: Ionic conductivity of single-crystal and polycrystalline $RbAg_4I_5$, *J. Appl. Phys.* 41 (1970) 1876.

[13] T. Takahashi, O. Yamamoto, S. Yamada, S. Hayashi: Solid-state ionics: High copper ion conductivity of the system CuCl–CuI–RbCl, *J. Electrochem. Soc.* 126 (1979) 1654–1658.

[14] W. van Gool: *Fast ion Transport in Solids: Solid State Batteries and Devices* (Elsevier Science Publishing Co. Inc., USA, 1973).

[15] Y.-F. Y. Yao, J. T. Kummer: Ion exchange properties and rates of ionic diffusion in beta-alumina, *J. Inorg. Nucl. Chem.* 29 (1967) 2453–2466.

[16] J. B. Goodenough, H. Y. P. Hong, J. A. Kafalas: Fast Na^+-ion transport in skeleton structures, *Mater. Res. Bull.* 11(2) (1976) 203–220.

[17] H. Y. P. Hong: Crystal structures and crystal chemistry in the system $Na_{1+x}Zr_2Si_xP_{3x}O_{12}$. *Mater. Res. Bull.* 11(2) (1976) 173–182.

[18] C. C. Liang, P. Bro: A High-Voltage, Solid-state battery system: I. Design considerations, *J. Electrochem. Soc.* 116(9) (1969) 1322–1323.

[19] C. C. Liang, J. Epstein, G. H. Boyle: A high-voltage, solid-state battery system: II. Fabrication of thin-film cells, *J. Electrochem. Soc.* 116(10) (1969) 1452–1454.

[20] B. J. H. Jackson, D. A. Young: Ionic conduction in pure and doped single-crystalline lithium iodide, *J. Phys. Chem. Solids* 30(8) (1969) 1973–1976.

[21] C. C. Liang: Conduction characteristics of the lithium iodide-aluminum oxide solid electrolytes, *J. Electrochem. Soc.* 120 (1973) 1289–1292.

[22] F. W. Poulsen, N. H. Andersen, B. Kindl, J. Schoonman: Properties of LiI-alumina composite electrolytes, *Solid State Ionics* 9–10 (1983) 119–122.

[23] J. B. Wagner: Composite solid ion conductors, in *High Conductivity Solid Ionic Conductors*, Edited by T. Takahashi (World Scientific, Singapore, 1989), pp. 146–165.

[24] J. Maier: Ionic conduction in space charge regions, *Prog. Solid State Chem.* 23 (1995) 171–263.

[25] A. A. Schneider, D. E. Harney, M. J. Harney: The lithium-iodine cell for medical and commercial applications, *J. Power Sources* 5(1980) 15–23.

[26] U. V. Alpen, A. Rabenau, G. H. Talat: Ionic conductivity in Li_3N single crystals, *Appl. Phys. Lett.* 30(1) (1977) 621.

[27] R. M. Yonco, E. Veleckis, V. A. Maroni: Solubility of nitrogen in liquid lithium and thermal decomposition of solid Li_3N, *J. Nucl. Mater.* 57(3) (1975) 317–324.

[28] A. Hooper, T. Lapp, S. Skaarup: Studies of hydrogen doped lithium nitride, *Mater. Res. Bull.* 14(12) (1979) 1617–1622.

[29] J. R. Rea, D. L. Foster, P. R. Mallory, I. Co: High ionic conductivity in densified polycrystalline lithium nitride, *Mater. Res. Bull.* 14(6) (1979) 841–846.

[30] M. Matsuo, S.-I. Orimo: Lithium fast-ionic conduction in complex hydrides: review and prospects, *Adv. Energy Mater.* 1(2) (2011) 161–172.

[31] M. Matsuo, Y. Nakamori, S.-I. Orimo, H. Maekawa, H. Takamura: Lithium superionic conduction in lithium borohydride accompanied by structural transition, *Appl. Phys. Lett.* 91(22) (2007) 224103.

[32] B. E. Taylor, A. D. English, T. Berzins: New solid ionic conductors, *Mater. Res. Bull.* 12 (1977) 171–181.

[33] H. Y. P. Hong: Crystal structure and ionic conductivity of $Li_{14}Zn(GeO_4)_4$ and other new Li^+ superionic conductor, *Mater. Res. Bull.* 13 (1978) 117–124.

[34] P. G. Bruce, A. R. West: Phase diagram of the LISICON, solid electrolyte system, Li_4GeO_4–Zn_2GeO_4, *Mater. Res. Bull.* 15 (1980) 379–385.

[35] Y. Inaguma, C. Liquan, M. Itoh, T. Nakamura, T. Uchida, H. Ikuta, M. Wakihara: High ionic-conductivity in lithium lanthanum titanate, *Solid State Comm.* 86(10) (1993) 689–693.

[36] V. Thangadurai, H. Kaack, W. Weppner: Novel fast lithium ion conduction in garnet-type $Li_5La_3M_2O_{12}$ (M = Nb, Ta), *J. Am. Ceram. Soc.* 86(3) (2003) 437–440.

[37] R. Murugan, V. Thangadurai, W. Weppner: Fast lithium ion conduction in garnet-type $Li_7La_3Zr_2O_{12}$, *Angew. Chem. -Int. Ed.* 46 (2007) 7778–7781.

[38] R. Kanno, T. Hata, Y. Kawamoto, M. Irie: Synthesis of a new lithium ionic conductor, thio-LISICON-lithium germanium sulfide system, *Solid State Ionics* 130 (2000) 97–104.

[39] R. Kanno, M. Maruyama: Lithium ionic conductor thio-LISICON: The Li_2S–GeS_2–P_2S_5 system, *J. Electrochem. Soc.* 148 (2001) A742–A746.

[40] M. Murayama, R. Kanno, M. Irie, S. Ito, T. Hata, N. Sonoyama, Y. Kawamoto: Synthesis of new lithium ionic conductor thio-LISICON-lithium silicon sulfides system, *J. Solid State Chem.* 168 (2002) 140–148.

[41] T. Kobayashi, Y. Imade, D. Shishihara, K. Homma, M. Nagao, R. Watanabe, T. Yokoi, A. Yamada, R. Kanno, T. Tatsumi: All solid-state battery with sulfur electrode and thio-LISICON electrolyte, *J. Power Sources* 182 (2008) 621–625.

[42] T. Inada, T. Kobayashi, N. Sonoyama, A. Yamada, S. Kondo, M. Nagao, R. Kanno: All solid-state sheet battery using lithium inorganic solid electrolyte, thio-LISICON, *J. Power Sources* 194 (2009) 1085–1088.

[43] T. Matsumura, K. Nakano, R. Kanno, A. Hirano, N. Imanishi, Y. Takeda: Nickel sulfides as a cathode for all-solid-state ceramic lithium batteries, *J. Power Sources* 174 (2007) 632–636.

[44] S. Boulineau, M. Courty, J.-M. Tarascon, V. Viallet: Mechanochemical synthesis of Li-argyrodite Li_6PS_5X ($X = Cl$, Br, I) as sulfur-based solid electrolytes for all solid state batteries application, *Solid State Ionics* 221 (2012) 1–5.

[45] N. Kamaya, K. Homma, Y. Yamakawa, M. Hirayama, R. Kanno, M. Yonemura, T. Kamiyama, Y. Kato, S. Hama, K. Kawamoto, A. Mitsui: A lithium superionic conductor, *Nat. Mater.* 10 (2011) 682–686.

[46] K. Otto: Electrical conductivity of SiO_2–B_2O_3 glasses containing lithium or sodium, *Phys. Chem. Glasses* 7(1) (1966) 29–37.

[47] A. Levasseur, J. C. Brethous, J. M. Reau, P. Hagenmuller: Comparative-study of ionic-conductivity of lithium in vitreous halogen borate, *Mater. Res. Bull.* 14(7) (1979) 921–927.

[48] A. Pradel, T. Pagnier, M. Ribes: Effect of rapid quenching on electrical properties of lithium conductive glasses, *Solid State Ionics* 17(2) (1985) 147–154.

[49] K. Tanaka, T. Yoko, H. Yamada, K. Kamiya: Structure and ionic conductivity of LiCl–Li_2O–TeO_2 glasses, *J. Non-Cryst. Solids* 103(2–3) (1988) 250–256.

[50] J. P. Malugani G. Robert: Ion conductivity in $LiPO_3$–LiX glasses (X = I, Br, Cl), *Mater. Res. Bull.* 14(8) (1979) 1075–1081.

[51] K. Tanaka, T. Yoko, K. Kamiya, H. Yamada, S. Sakka: Properties of oxybromide tellurite glasses in the system LiBr–Li_2O–TeO_2, *J. Non-Cryst. Solids* 135(2–3)(1991) 211–218.

[52] J. B. Bates, N. J. Dudney, G. R. Gruzalski, R. A. Zuhr, A. Choudhury, C. F. Luck, J. D. Robertson: Electrical properties of amorphous lithium electrolyte thin films, *Solid State Ionics* 53–56 (1992) 647–654.

[53] J. B. Bates, N. J. Dudney, G. R. Gruzalski, R. A. Zuhr, A. Choudhury, C. F. Luck: Fabrication and characterization of amorphous lithium electrolyte thin films and rechargeable thin-film batteries *J. Power Sources* 43(1–3) (1993) 103–110.

[54] J. B. Bates, N. J. Dudney, B. Neudecker, A. Ueda, C. D. Evans: Thin-film lithium and lithium-ion batteries, *Solid State Ionics* 135(1–4) (2000) 33–45.

[55] http://www.cymbet.com/.

[56] R. Mercier, J.-P. Malugani, B. Fahys, G. Robert: Superionic conduction in Li_2S-P_2S_5-LiI-glasses, *Solid State Ionics* 5 (1981) 663–666.

[57] H. Wada, M. Menetrier, A. Levasseur, P. Hagenmuller: Preparation and ionic conductivity of new B_2S_3–Li_2S–LiI glasses, *Mater. Res. Bull.* 18 (1983) 189–193.

[58] J. H. Kennedy, Y. Yang: A highly conductive Li^+-glass system: $(1\text{-}x)\cdot$ $(0.4SiS_2\text{-}0.6Li_2S)$-$x$LiI, *J. Electrochem. Soc.* 133 (1986) 2437–2438.

[59] F. Mizuno, A. Hayashi, K. Tadanaga, M. Tatsumisago: New, highly ion-conductive crystals precipitated from Li_2S-P_2S_5 glasses, *Adv. Mater.* 17 (2005) 918–921.

Chapter 4

Ion Conduction in Ceramics

4.1 Structural Fundamentals of Ion Conductive Ceramics

Solid electrolytes, i.e. solid-state ion conductors, consist of mobile ions and a rigid skeleton composed of a polyhedral network of metal or nonmetal ions with ligands (normally anions). High ionic conductivity often conflicts with high structural stability, because higher ionic conductivity requires weak chemical bonding energy, which leads to opposite effects on the structural stability [1].

More than half the elements in the periodic table are related to the solid-state ion conductors [2]. Some of cations and anions such as H^+, Li^+, Na^+, K^+, Mg^{2+}, Cu^+, Ag^+, F^-, Cl^- and O^{2-} have been found to be mobile in the solid state (Fig. 4.1). Many metal and nonmetal ions coordinated by ligands (chalcogens, halogens and nitrogen) have formed polyhedral networks. Early transition-metal ions in the first and second rows, such as Ti^{4+}, Zr^{4+}, Nb^{5+}, or Ta^{5+}, have been used very often because they have no electron in the d-orbital (their outer electron structures are $s^2p^6d^0$). Accordingly, these ions do not have the significant electronic conductivity that makes the solid electrolytes change from an ion conductor to a mixed conductor, which possesses both ionic and electronic conductivities. Additionally, ions in groups 13 (e.g. Al^{3+} and Ga^{3+}), 14 (e.g. Si^{4+} and Ge^{4+}) and 15 (e.g. P^{5+}) are also used to form polyhedra in 12-fold, 8-fold, 6-fold or 4-fold coordination with the ligand. The polyhedral

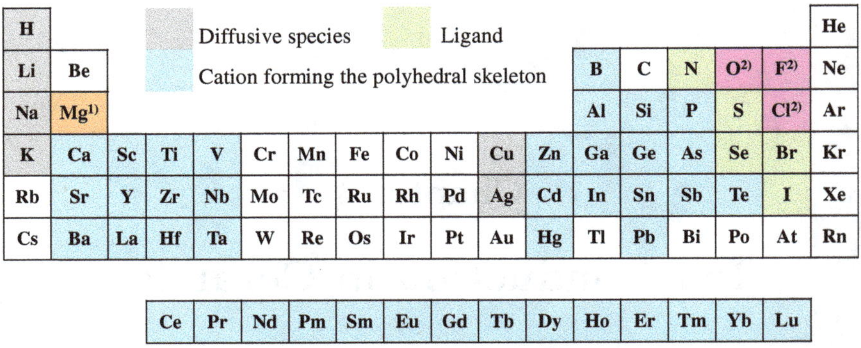

H			Diffusive species		Ligand												He
Li	Be		Cation forming the polyhedral skeleton									B	C	N	O[2]	F[2]	Ne
Na	Mg[1]											Al	Si	P	S	Cl[2]	Ar
K	Ca	Sc	Ti	V	Cr	Mn	Fe	Co	Ni	Cu	Zn	Ga	Ge	As	Se	Br	Kr
Rb	Sr	Y	Zr	Nb	Mo	Tc	Ru	Rh	Pd	Ag	Cd	In	Sn	Sb	Te	I	Xe
Cs	Ba	La	Hf	Ta	W	Re	Os	Ir	Pt	Au	Hg	Tl	Pb	Bi	Po	At	Rn

Ce	Pr	Nd	Pm	Sm	Eu	Gd	Tb	Dy	Ho	Er	Tm	Yb	Lu

Fig. 4.1   Elements used in ion conductive ceramics. (1) Mg is used as a diffusive specie or forms polyhedral skeleton. (2) O, F and Cl work as diffusive species or ligands.

Source: Reproduced from Ref. [2] with permission from Royal Society of Chemistry.

networks exist even in the amorphous solid electrolyte. The difference is only whether the solid electrolyte has long-range ordering of the polyhedron or not. The polyhedra are connected with each other by corner-sharing like in NASICON and edge/corner-sharing like in garnets to form an ion conduction path.

The ion conduction in crystals is significantly different from that in the liquid electrolyte. In the liquid electrolyte, the lithium ion is solvated [3], and the solvated Li ion moves from the positive electrode to the negative electrode and vice versa. The Li ion conductivity in the liquid electrolyte relates to the Stokes–Einstein equation and can be enhanced by increasing Li ion concentration using a high concentration of Li salts. The usage of the liquid electrolyte with high dielectric constant promotes ion dissociation and lowers the viscosity of the liquid electrolyte, enhancing the Li ion mobility. The potential energy profile of the mobile Li ion in the liquid electrolyte can be considered to be flat because of uniform surrounding (Fig. 4.2(a)).

On the contrary, conduction of Li ion in the crystals is through diffusion of ions passing through periodic bottleneck points in the polyhedral network. This means the mobile ion must overcome the energy barriers at the bottlenecks that appear periodically

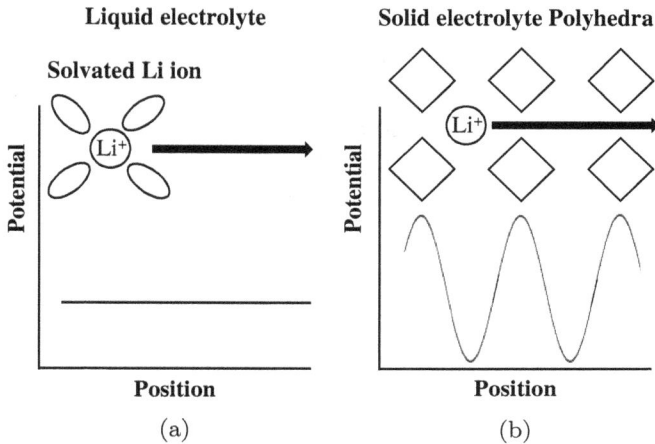

Fig. 4.2 Potential change of Li ion conduction in (a) liquid electrolyte and (b) ceramics.

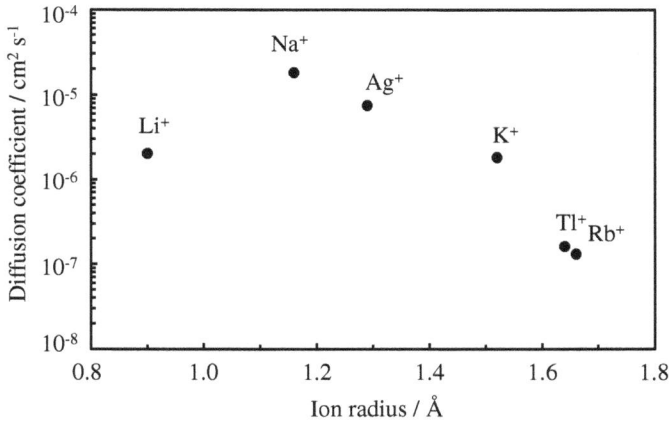

Fig. 4.3 Diffusion coefficient of monovalent cations in substituted β-Al$_2$O$_3$ at 400°C [4].

(Fig. 4.2(b)). As can be expected, a relationship between the bottleneck and mobile ion sizes significantly influences the ion conduction. Figure 4.3 shows diffusion constant as a function of ion radius in substituted β-Al$_2$O$_3$ at 400°C [4]. As can be seen, Na$^+$ ion leads to maximum diffusion constant. This implies that the highest conductivity can be obtained in ions that are of a suitable size for a given structure. When the mobile cation is too small, the cation occupies a site with a large electrostatic well, which makes the neighboring counterions come closer, resulting in high activation energies and slow diffusion. On the other hand, when the mobile cation is too large, the cation experiences larger forces when it passes through the bottlenecks of the skeleton structure, leading to reduced diffusivities and large activation energies.

Therefore, optimization of bottleneck size by aliovalent substitution is a good and common strategy to improve the Li ion conductivity. Apart from the ion size, ion vacancy also greatly influences the ion conductivity. Ionic conductivity and diffusivity decrease with increasing ion vacancy due to the increased electrostatic interactions between mobile ions and counterions forming the structural skeleton. Figure 4.4 depicts dependency of diffusion coefficient on monovalent, divalent and trivalent ions in aliovalent and isovalent substituted lithium sulfates at 550°C [5]. The diffusion coefficient can decrease by 3 orders of magnitude from monovalent to trivalent ions. This is the reason that the highest conductivity is obtained in monovalent ion conductors such as silver, lithium and sodium.

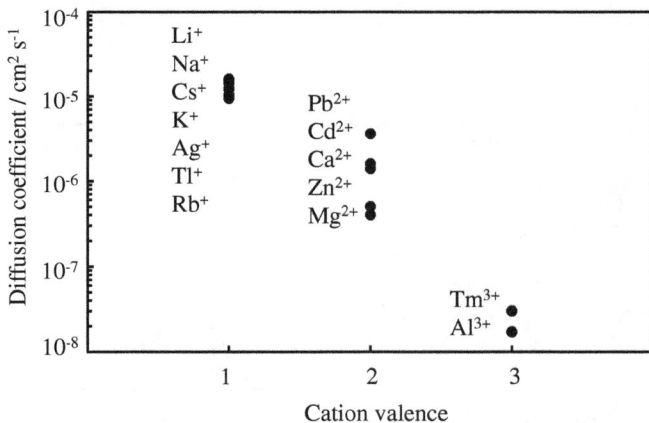

Fig. 4.4   Diffusion coefficient of monovalent, divalent and trivalent cations in Li_2SO_4 at 550°C [5].

4.2 Ion Conduction Inside the Crystal

The ionic conduction in the crystal not only depends on a rigorous correlation between mobile species and structure but is also caused by defects, e.g. vacancies, interstitials and partial occupancy on lattice sites or interstices. The formation of these disorders in stoichiometric ion conductors is determined by the ionic gap or defect formation energy (E_f). Moreover, defects can also be created by doping and substitution of elements with different valence states whose formation energetics is governed by the trapping energy (E_t). In the highly disordered structure, the sites available for ion movement are much larger than the number of the mobile ions themselves, thereby leading to high ionic conductivity. Meanwhile, introduction of foreign heteroatoms usually tunes the structural parameters such as lattice constants and unit cell volume and further facilitates the ionic conduction. These two kinds of defect formation energies, E_f and E_t, contribute to the apparent activation energy for ion diffusion and migration. A more detailed explanation of defect is given in the next section.

4.2.1 *Defect*

When the crystal is perfectly ordered (Fig. 4.5(a)), the ions cannot leave their lattice sites. Once thermodynamics-driven defects are formed, the ions can diffuse through the defects (Fig. 4.5(b)). The super-ion conductor possesses a low energy barrier that the ions must overcome to move to adjacent defects (vacancies). To exhibit these defects, the commonly used nomenclature was defined by Kroeger and Vink [6]. Point defects are considered to be similar to a solution in which the salt (defect) is diluted by a solvent (solid). Several similarities can be found between defect formation in the solids and salt dissolving in the solvent. These are as follows: (1) formation of a pair of charged defects which relate to electrical conduction and (2) defect concentration is influenced by temperature (defects are thermally activated). In the Kroeger–Vink notation, a dot ($\bullet$), an x-mark ($\times$) and a prime ($'$) in superscript denote positive, neutral and negative charge, respectively. The subscript indicates the site of

Defect free structure   Defect containing structure

(a)   (b)

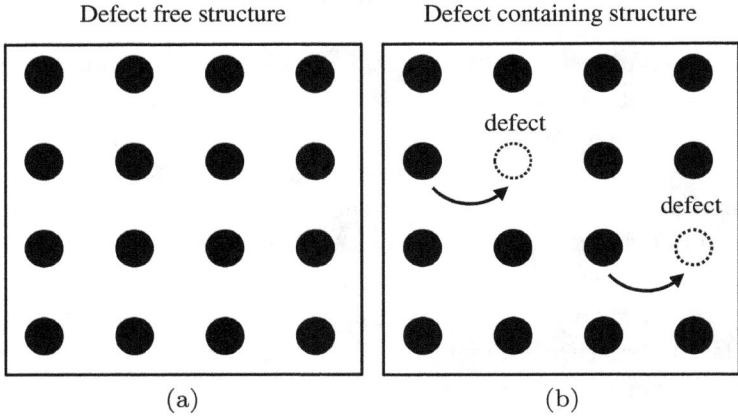

Fig. 4.5   (a) Defect-free structure and (b) defect containing structure.

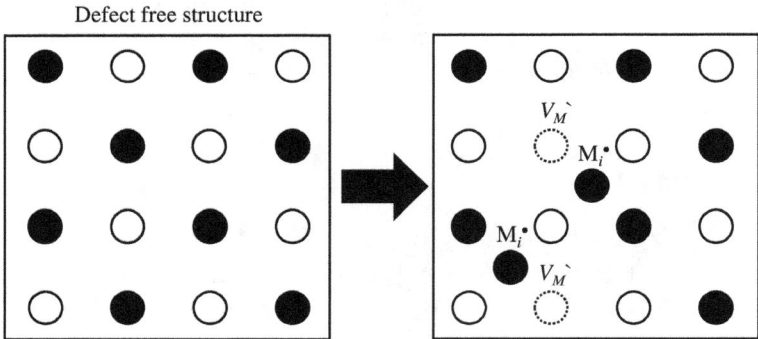

Defect free structure

Fig. 4.6   Frenkel defect.

a defect, e.g. the subscript i represents an interstitial site. Vacancy is denoted by V. For example, $M_i^\bullet$ means an interstitial metal ion which is charged positively and $Vo^{\bullet\bullet}$ represents a doubly charged oxygen vacancy. Bulk defect chemical reactions must obey mass balance, charge balance (electrical neutrality) and lattice-site balance.

The formation of an interstitial positively charged metal ion is described by Frenkel reaction (Fig. 4.6).

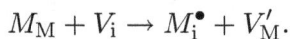

$$M_M + V_i \rightarrow M_i^\bullet + V_M'.$$

It can be imagined that the interstitial formation is preferred in small ions and/or open lattices. In fact, cationic Frenkel defect is more often observed than the corresponding anionic defect (called anti-Frenkel type), because anions are generally larger than cations. Especially, ionic solids with large size difference between cations and anions, such as ZnS, AgCl and AgI, usually exhibit the Frenkel defect. The Frenkel defect affects the electrical conductivity of the solids.

When defects are formed by the displacement of a cation–anion pair, it is called Schottky defect (Fig. 4.7). For example,

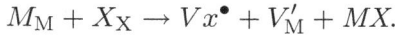

$$M_M + X_X \rightarrow Vx^\bullet + V'_M + MX.$$

Here, MX represents ions that have been displaced to new sites. This type of defect is observed mainly in dense crystal lattices such as NaCl and RbI. The Schottky defect does not change the electrical conductivity, but the density of material.

Defects can also be created by changing stoichiometry through heteroatom substitution or formation of impurities. This induces changes of chemical potential and formation of ionic and electronic charge carriers. For example, Y-stabilized ZrO_2 (YSZ) which contains Y^{3+} and Zr^{4+} possesses defects in the oxygen site to compensate the

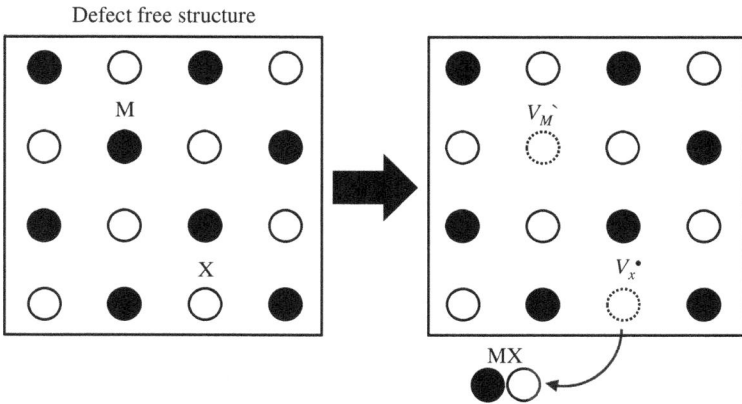

Fig. 4.7   Schottky defect.

lower cation valence of Y^{3+} than Zr^{4+} [7].

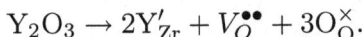

$$Y_2O_3 \rightarrow 2Y'_{Zr} + V_O^{\bullet\bullet} + 3O_O^{\times}.$$

The oxygen vacancy (defect) works as an oxygen carrier.

More simply, oxidation or reduction of materials also sometimes causes deviation from stoichiometry and creates ionic and electronic carriers. Electron holes and oxygen vacancies are formed by oxidation and reduction, respectively

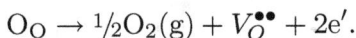

$$\tfrac{1}{2}O_2(g) + V_O^{\bullet\bullet} \rightarrow O_O + 2h^{\bullet},$$

$$O_O \rightarrow \tfrac{1}{2}O_2(g) + V_O^{\bullet\bullet} + 2e'.$$

The concentration of defects affects the ionic conductivity. The concentration of both Schottky and Frenkel defects can be expressed in terms of thermodynamics parameters under different conditions.

Taking Schottky defect as an example, when the mole fractions of positive and negative vacancies are expressed as x_1 and x_2 and their numbers as n_+ and n_-, respectively, and N is the number of cation or anion sites, the charge neutrality in a pure crystal can be written as:

$$x_1 = x_2 = x_0, \tag{4.1}$$

$$n_+ = n_- = N \exp(-G_s/kT), \tag{4.2}$$

$$x_1 x_2 = x_0^2 = \exp(-G_s/kT) = \exp(S_s/k)\exp(-H_s/kT), \tag{4.3}$$

where G_s, S_s and H_s represent the Gibbs energy, entropy energy and enthalpy energy of the Schottky defect formation, respectively.

Likewise, the Frenkel defects can expressed as

$$n_F = (NN')^{1/2} \exp(-G_F/2kT). \tag{4.4}$$

Here, N' is the number of interstitial cites and G_F is Gibbs energy of the Frenkel defect formation.

Considering the case of cation doping, e.g. Ca^{2+} in $NaCl$, there will be additional vacancies produced to balance the charge difference in the structure through doping. Then, the charge neutrality can be written as

$$x_1 = x_2 + c_1, \tag{4.5}$$

where c_1 is the mole fraction of cation impurities.

Accordingly, the mole fractions can be written as

$$x_1 = 1/2c_1\{[1 + (2x_0/c_1)^2]^{1/2} + 1\}, \tag{4.6}$$

$$x_2 = 1/2c_1\{[1 + (2x_0/c_1)^2]^{1/2} - 1\}. \tag{4.7}$$

If the amount of cations is relatively small, $c \leq x_0$, the vacancy concentrations are almost equal to that of the pure crystal, i.e. $x_1 = x_2 = x_0$; while for large amount of doping, i.e. $c \geq x_0$, the vacancy concentrations would be

$$x_1 = c_1 \quad \text{and} \quad x_2 = x_0^2/c_1. \tag{4.8}$$

Likewise, for a divalent anion doped with a concentration c_2, the neutrality conditions become

$$x_2 = x_1 + c_2 \tag{4.9}$$

and the values of x_1 and x_2 would be

$$x_1 = 1/2c_2\{[1 + (2x_0/c_2)^2]^{1/2} - 1\}, \tag{4.10}$$

$$x_2 = 1/2c_2\{[1 + (2x_0/c_2)^2]^{1/2} + 1\}. \tag{4.11}$$

4.2.2 *Hopping model*

In ordinary solids, mobile ions are in perpetual movement in all possible directions, from one lattice point to the other through vacancy and interstitial sites. Because of this thermally activated random motion, the concentration of defects and ions inside the lattice are uniform throughout the solid, and this is referred to as diffusion. In ceramic electrolytes, the high conductivity is the product of a combination of a high concentration of mobile species and a low energy bottleneck for ion motion from one site to another. In the presence of electric field, Li^+ ions would still move randomly, while also migrating toward the direction of the electric field, which constitutes the basic process of Li^+ conduction in crystalline solids.

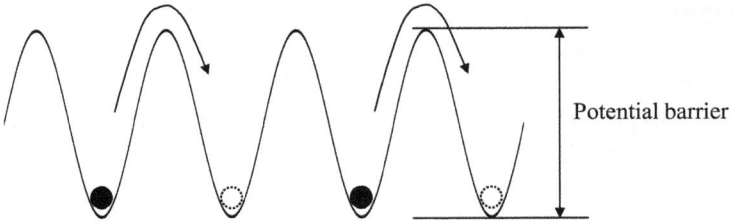

Fig. 4.8   Ion hopping model.

Li^+ ion transport mechanism in ceramic materials is a complicated process. To deepen our understanding, quite a few theoretical and experimental models have been built, e.g. hopping model, two-state model, lattice gas model, frequency methods, etc. These models help to shed light on this phenomenon.

The hopping model regards ionic conduction as a jumping process through the lattice. In this model, ion conduction occurs by ion hopping from one site to vicinity site (Fig. 4.8), which is based upon two factors:

(a) The probability of Li^+ ion jumping to an adjacent site in a given direction in a unit time, which is referred as jumping frequency.
(b) The probability that a given site has available neighbor sites for ion conduction, i.e. the product of defects concentration and the number of nearest neighbor sites.

The driving force of hopping is thermal activation [8, 9]. The diffusion coefficient of the ion, D_i, is expressed as a function of jump distance (a), the attempt frequency, ν_0, and the free energy of migration ($\Delta G_{mig} = \Delta H_{mig} - T\Delta S_{mig}$) by Boltzmann statistics

$$D_i = \gamma a^2 \nu_0 \exp(-\Delta G_m/\kappa T). \qquad (4.12)$$

The factor γ is used to take into account geometrical and correlation effects. From the Nernst–Einstein equation, the mobility of ion, μ_i, can be described as

$$\mu_i = D_i q_i/\kappa T, \qquad (4.13)$$

where q_i, κ and T are the charge of the ions, the Boltzmann constant and the absolute temperature, respectively. Also, ion conductivity is shown in Eq. (4.14).

$$\sigma = q_i \mu_i c_i, \qquad (4.14)$$

where c_i is the density of carrier.

From Eqs. (4.12)–(4.14), the ionic conductivity can be described as

$$\sigma = (q_i^2 c_i / \kappa T) \times \gamma a^2 \nu_0 \exp(-\Delta H_m / \kappa T) \exp(\Delta S_m / \kappa). \qquad (4.15)$$

The density of the carrier is a function of temperature because the carrier can be thermally activated.

$$c_i = c_{i0} \exp(-\Delta H_f / 2\kappa T), \qquad (4.16)$$

where H_f is the formation energy of the carrier.

Accordingly, the ionic conductivity is written as:

$$\sigma = (\sigma_0 / T) \exp(-\Delta H_a / \kappa T), \qquad (4.17)$$

where ΔH_a is the sum of formation and migration enthalpies ($\Delta H_a = \Delta H_m + \Delta H_f / 2$).

The above expressions are applied for the case of thermodynamic equilibrium where the probability of jumping in each direction is equal. While in the presence of an imposed electric field, a potential gradient is generated throughout the solid. Therefore, Li ions would experience an asymmetric potential energy barrier when jumping from site to site. In this case, due to the applied electric field E, an additional term, $(a/2)qE$, needs to be accounted for in the potential energy.

For a jump corresponding to the direction of the applied field, the probability would therefore be higher with decreased potential energy

$$\Delta G_m - (a/2)qE.$$

For a jump against the field, it occurs with reduced probability due to the higher potential energy

$$\Delta G_m + (a/2)\,qE.$$

It should be noted that not only the Li ion conductors but also most crystalline and amorphous inorganic ion conductors follow Eq. (4.17). Therefore, it has been considered that ion conduction inside the crystal is caused by ion hopping.

4.3   Ion Conduction at the Interface

Among crystalline materials, polycrystalline materials are usually used because of the difficulty in preparing a single crystal. In this case, two ion conduction pathways, i.e. conduction inside the crystal (bulk conduction) and at the interface (grain boundary conduction), must be considered (Fig. 4.9) [10].

When ion conductive ceramics are used for electrochemical devices, total ion conductivity, which is composed of bulk conductivity and grain boundary conductivity, is important. The ion conduction inside the crystal grains is well explained by the ion hopping model, while the ion conduction at the crystal grain boundary of Li ion conductive ceramics is usually inferior to that inside the crystal grain. The ion conduction at the grain boundary becomes a rate-determining step of the Li ion conduction in many cases. At the interface of crystal grains, discontinuity of Li ion conduction path

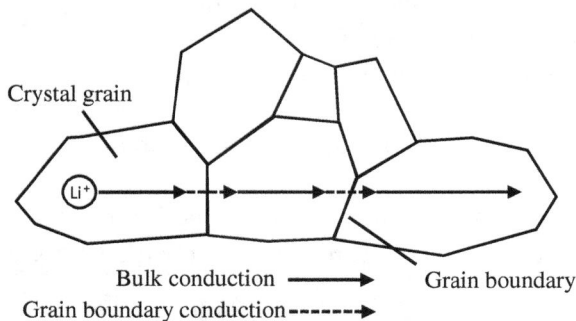

Fig. 4.9   Ion conduction in a polycrystalline material.

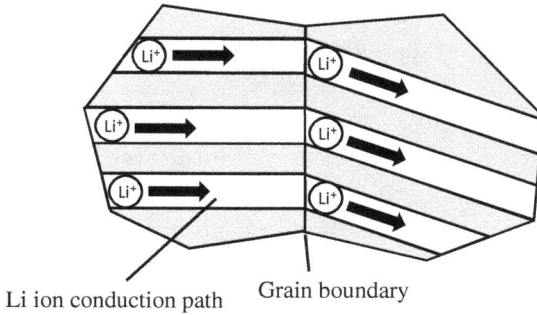

Li ion conduction path   Grain boundary

Fig. 4.10   Li ion conduction at grain boundary.

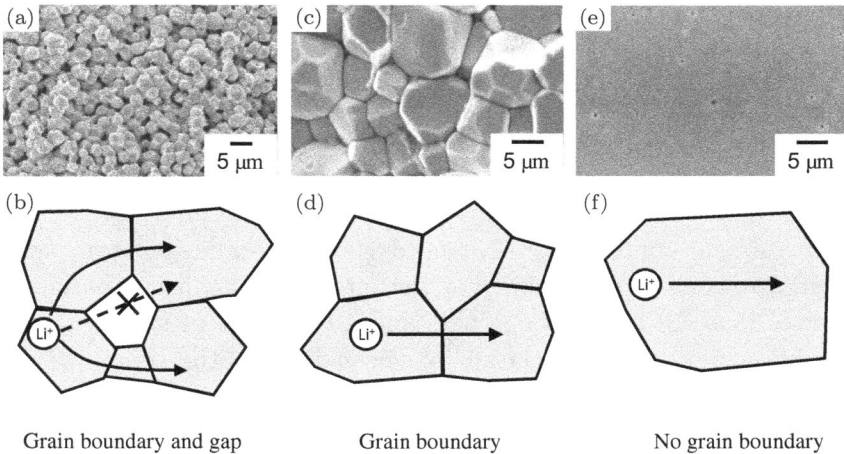

Grain boundary and gap   Grain boundary   No grain boundary

Fig. 4.11   Illustration of grain boundary and void vs transportation of Li ions: (a), (c) and (e) SEM images of ceramics with different grain sizes, and (b), (d) and (f) schematic illustration of ion conduction in the ceramics with different grain sizes.

impedes the Li ion transportation, as schematically illustrated in Fig. 4.10. In the sample with low sinterability, gaps exist among the crystal grains (Figs. 4.11(a) and 4.11(b)). Li ion cannot migrate in the gaps, and thus the gaps narrow the Li ion conduction path, lowering the grain boundary conductivity. This leads to low total ion conductivity, even though the bulk conductivity is high. With increase of the sinterability of sample, the gaps disappear, leading to increased Li ion conduction pathway (Figs. 4.11(c) and 4.11(d)). The grain boundaries also disappear with further increase of sinterability (Figs. 4.11(e) and 4.11(f)). In this case, the total conductivity is consistent with the bulk conductivity. Additionally, low conductive impurities tend to accumulate at the interface. These likely work as the gaps in the low sinterability sample, lowering the ion conductivity. On the other hand, the impurities improve the ion conductivity in some cases because of the space charge and percolation effects. The space charge effect also can be applied for an interface between electrode materials and the solid electrolyte.

4.3.1   *Space charge effect*

The interfaces between materials, such as grain boundaries, surface and electrode/electrolyte interface, include a structural and/or compositional discontinuity that is accompanied by changing charge-carrier concentration. Ion conductivity is influenced by electrostatic charge that presents at the interface [11]. On the interface between two different materials, unbalanced charges would generate space charge regions in the vicinity of the interface with accumulation of mobile ions (Li ions) on one side and a depletion of mobile species (in other words, accumulation of vacancies) on the other [12–15]. Space charge effects have been known in colloid systems, in which the space charge concept has been used to explain conductivity effects in two-phase materials [16]. This space-charge theory for ion conductive materials is further refined and extended [11]. Briefly, it can be summarized as follows. Ion defects will form and be trapped at the interface and counterdefects will accumulate in the adjacent space charge region. The driving force for trapping the defects is attributed to the presence of a second phase which has a chemical affinity for the mobile ion. For example, there are many nucleophilic hydroxide surface groups on the basic oxides. The surface groups can attract and fix cations [17]. The gas phases can also form space charge region on the surface of solid materials. For example, NH_3 can form complexes with silver or copper ions in aqueous solution. Therefore, silver and copper ions can be trapped on the ion conductive material

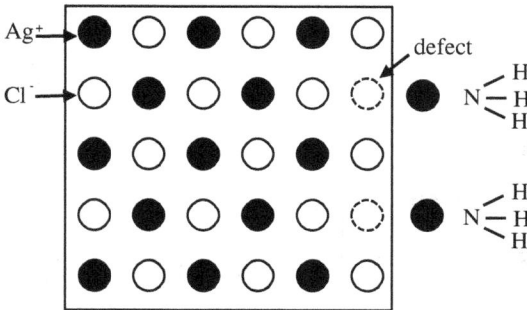

Fig. 4.12   Creation of Frenkel defect by NH_3.

surface when it contacts a gas phase containing NH_3 (Fig. 4.12) [18, 19].

The defect reaction in a Frenkel-disordered material M^+X^- can be expressed as

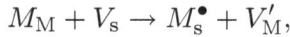

$$M_M + V_s \rightarrow M_s^{\bullet} + V_M',$$

where the main symbol M represents ion and V represents vacancy. The subscripts and superscripts represent the site of lattice and type of charge, respectively. For example, $M_S^{\bullet}$ means a metal ion in the position "S" that is positively charged, while V_M' means a vacancy at "M" where it is negatively charged. Figure 4.13 schematically shows vacancies that are distributed in the adjacent interface region, which forms an electrically charged space charge region (Fig. 4.13).

Although the profiles of space charge region are different between solid–gas and solid–solid contacts, the thickness of space charge layer is approximately twice the Debye length in both cases [11]. The defect concentration in bulk is a function of temperature, chemical potential and dopants, while the defect concentration in the space charge layer depends on the difference of chemical potential between the bulk and space charge layer [11]. This space charge layer effect has been observed, for example, in $Li_{1.5}Al_{0.5}Ge_{1.5}(PO_4)_3$ (LAGP) glass ceramics containing $AlPO_4$ [20]. There is a reflection point in the Arrhenius plot in terms of Li ion conductivity of the sample as shown in Fig. 4.14. The adsorption of Li ion onto a dielectric $AlPO_4$ particle surface at the grain boundary caused the space charge effect,

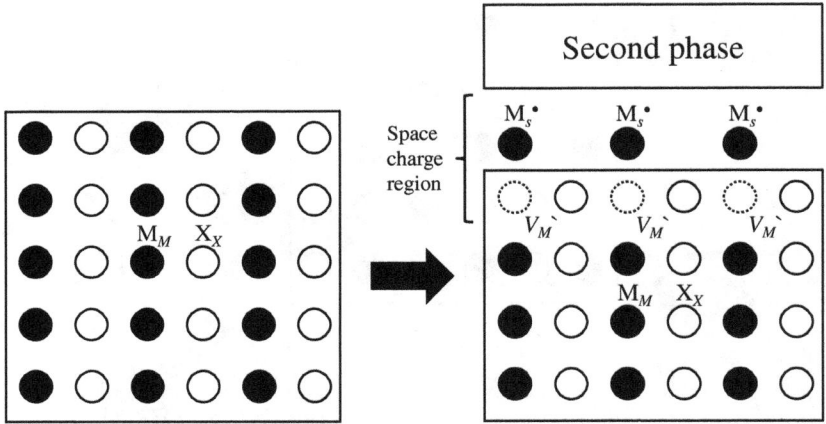

Fig. 4.13 Space charge effect of Frenkel-disordered material.

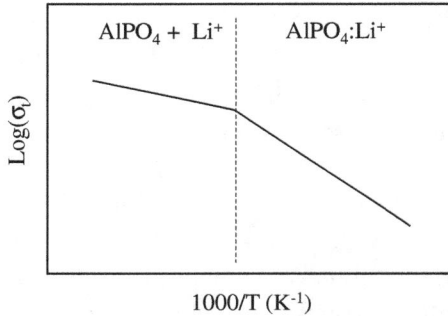

Fig. 4.14 Arrhenius plot of LAGP containing $AlPO_4$.

which is schematically shown in Eq. (4.18).

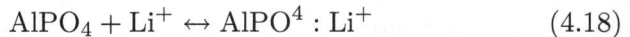

$$AlPO_4 + Li^+ \leftrightarrow AlPO^4 : Li^+ \qquad (4.18)$$

This $AlPO_4$:Li^+ complex mediated the Li ion conduction. The $AlPO_4$:Li^+ complex was decomposed above 37°C and the space charge effect disappeared, resulting in the observed reflection point in the Arrhenius plot. As expected, the particle size and distribution of $AlPO_4$ influence the Li ion transport of the sample. The sample with shorter crystallization time shows high Li ion conductivity by the space charge effect in fewer number and smaller in size of $AlPO_4$ particles. With increase of the crystallization time, the $AlPO_4$ particles become larger and more abundant. In this situation, the blocking effect of insulative $AlPO_4$ particle is dominant, leading to a lowering of the Li ion conductivity.

In a thin film where the film thickness is similar to the Debye length, an additional increase of conductivity can be expected. The thickness of the space charge region is about twice the Debye length; thus, defect concentration no longer reaches bulk value [11]. In this case, local charge neutrality is not at all satisfied. A full depletion (or accumulation) of charge carriers leads to enhancement of ion and/or electron conductivities. This concept is well supported by the study that the ionic conductivity of intrinsic epitaxial BaF_2/CaF_2 multilayers could be enhanced when layers become thinner, as shown in Fig. 4.15 [2]. With thinner layers, the relative volume affected by space charge layer becomes quite large, rendering higher concentration of charge carriers, thereby enhancing the ionic conductivities.

However, the space charge layer would also seriously limit the performance of solid electrolytes in lithium ion battery application. Generally, if a sulfide-based electrolyte is used, issues like lattice mismatch or poor contact should not be a significant problem as sulfides have good ductility and could easily build a good interfacial contact by simply cold pressing. Nevertheless, as proposed by Takada's research group, large interfacial resistance still exists, even though the sulfide-based electrolytes have comparable ionic conductivity as those of liquid electrolytes [21, 22]. This common phenomenon was then explained by the space charge layer effect, suggesting that the chemical potential difference between the oxide cathode and the sulfide electrolyte leads to the lithium ion depletion on the sulfide side, thereby forming a reduced Li-ion concentration interlayer with high resistance. To solve this problem, an interposed buffer layer is usually required to shield the direct contact between sulfide electrolyte and high potential oxide cathode. This buffer layer should have the following characteristics: a material with lithium

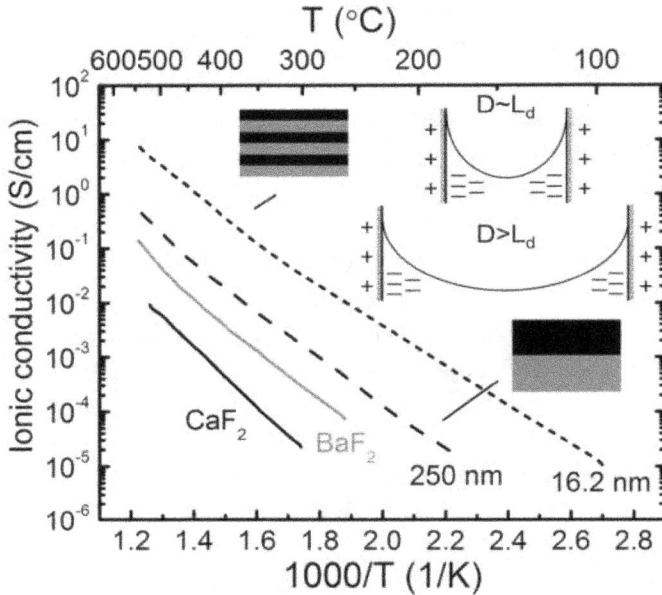

Fig. 4.15   Conductivity as a function of temperature for BaF_2–CaF_2 super lattices (dotted lines) as well as bulk BaF_2 and CaF_2 (solid lines). (Inset) Space charge layers for distances (D) between interfaces on the order of the Debye length (L_d) and for distances.

Source: Reproduced from Ref. [2] with permission from Royal Society of Chemistry.

ion conductance but with electron insulation. Takada *et al.* investigated various coating materials, including $LiNbO_3$ [23], $LiTaO_3$ [21] and $Li_4Ti_5O_{12}$ [24], and confirmed that an interposition of these materials with appropriate thickness could largely reduce the interfacial impedance. Density functional theory (DFT) simulation also confirms the interface equilibrium under various conditions. Figure 4.16 compares the interfacial Li concentration with or without $LiNbO_3$ interlayer by DFT calculations [25].

4.3.2   *Percolation effect*

The ion conductivity of composite electrolytes, which are composed of ion conductor and insulator, sometimes increases with the content

Conventional: Equilibrium

Calculation: Equilibrium

Calculation: Initial Stage of Charging

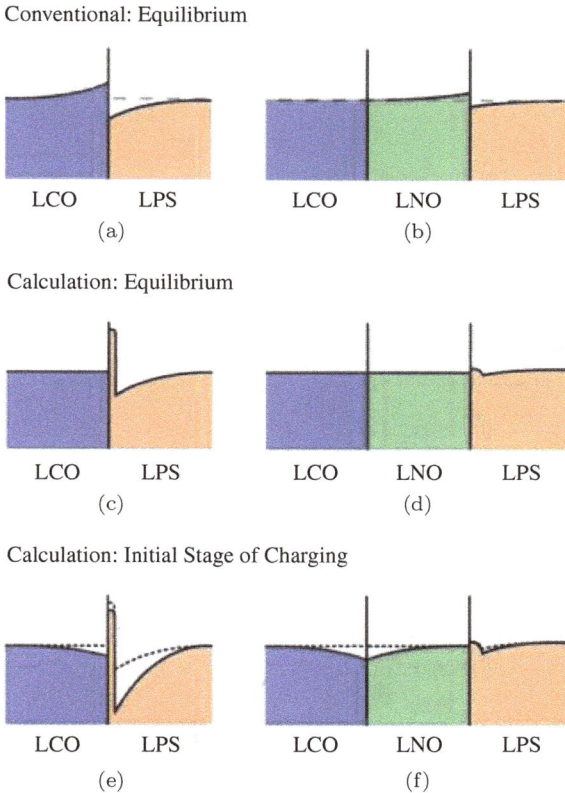

Fig. 4.16 Schematic illustrations of interfacial Li concentration with or without LiNbO$_3$ interlayer [25].

of the insulator and then reaches a maximum value. After that, the conductivity drops sharply with further increase of insulator content. Figure 4.17 shows the Li ion conductivity of 0.1M LiClO$_4$-methanol and 0.1M LiClO$_4$–THF (tetrahydrofuran) solutions containing various fractions of SiO$_2$ (ϕ) [26]. The conductivity was constant at low ϕ and then increased abruptly with ϕ. The maximum conductivity appeared at around $\phi = 0.2$. With further increase of ϕ, the conductivity dropped sharply and reached zero. This behavior can be explained by the percolation effect. The composite is considered to consist of three phases, ion conductive phase (methanol or THF

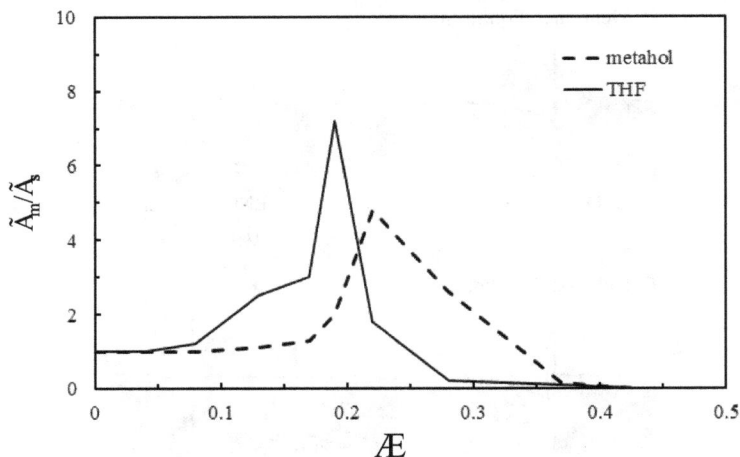

Fig. 4.17   Conductivity of 0.1M LiClO$_4$–methanol and 0.1M LiClO$_4$–THF with various fractions of SiO$_2$ (φ) [26].

solutions), insulator phase (SiO$_2$) and interphase, which has very high conductivity because of high concentration of charge carrier in the space charge region. For a small concentration of the insulator phase, the high conductive phases on the insulator particles are isolated from each other in the matrix of the ion conductive phase. Accordingly, the high conductive phases do not effectively contribute to ion conductivity enhancement (Fig. 4.18(a)). This corresponds to the constant conductivity region in Fig. 4.17. When the insulator concentration reaches a critical concentration, a continuous network of highly conducting paths extends through the whole sample (Fig. 4.18(b)). In this region, enhancement of the conductivity with ϕ was observed. With further increase of the insulator concentration, a second critical concentration occurs, in which high conduction paths on the insulator particles are partially disrupted, because continuous layers of insulator grains are formed (conductor–insulator transition) (Fig. 4.18(c)). The conductivity decreases sharply after this second threshold (Fig. 4.18(d)). The size of insulator particle also influences the conductivity significantly [26]. For example, when the particle size of the insulator is large and the thickness of high conductive

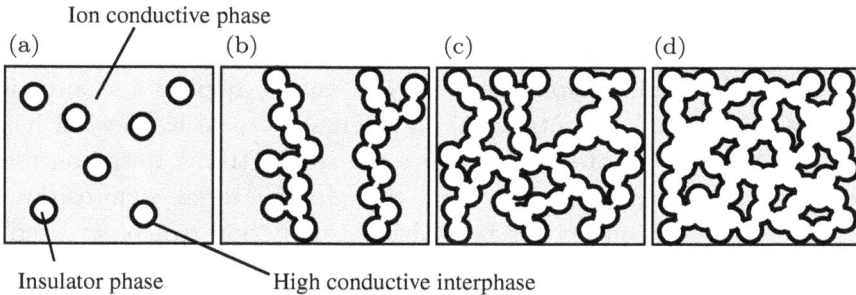

Fig. 4.18 Percolation model of insulator-dispersed ion conductor at different insulator concentrations. High conductive interphase is formed on the insulator surface by the space charge effect. (a) Low concentration where the interphase is isolated, (b) first critical concentration where one interphase connects with the other, (c) second critical concentration where the interphase network is partially collapsed and (d) high concentration of the insulator where the interphase network is collapsed [25].

interphase layers on the particles is negligible as compared with the particle size of insulator, it is very hard to obtain a continuous network of high conductive paths. The enhancement of conductivity cannot be observed. On the contrary, if the size of insulator is small and comparable to the thickness of interphase layer, the high conductive interphases can form a continuous network in the whole sample, resulting in an increase of ion conductivity. An abnormal enhancement of Li ion conductivity was observed by addition of Al_2O_3 to LiI [27]. LiI containing 33–45 mol% Al_2O_3 exhibited 50 times higher Li ion conductivity than pristine LiI. This is a good example of the space charge and percolation effects.

4.4 Strategies for Lithium Conductivity Enhancement

Intensive efforts have been made to enhance the conducting performance of solid electrolytes. These efforts are mostly focused on increasing Li ion concentration, creating more available sites and designing an open structure for Li ion movement. Generally, there are two strategies for lithium conductivity enhancement: compositional engineering and morphological engineering. These two strategies are sometimes applied simultaneously to improve the total conductivity by tuning bulk conductivity and grain boundary conductivity.

From the compositional engineering aspect, doping and substituting by various elements within a given structural framework has been an effective strategy to enhance the conductivity. Based on the knowledge of defect chemistry, by changing material composition, more vacancies required for fast Li ion conduction can be created. High lithium concentration can also be achieved by substitution of low valence elements. In addition, the structural framework can usually be finely tuned due to the different ionic radii of dopants, and therefore help further optimize lattice volume and bottleneck sizes for ionic movement. Although this is an effective strategy to design fast ion conductors, it is not straightforward to find the most conducting material by this method. Most of the existing fast ion conductors have been found through time-consuming trial and error techniques. Within each structural family, the lithium ion conductivity can vary greatly by 5 or 6 orders of magnitude. Thus, in order to better understand the relationship between Li ion transport mechanisms and conducting performance, much more work is needed, and the method of high-throughput calculations is also required to provide insights into this field.

On the other hand, from the morphological engineering aspect, the most common way is applying glass–ceramic state rather than crystalline state to reduce grain boundary resistance, especially for conductors with extremely low grain boundary conductivity. The OHARA company has already commercialized the NASICON (Na super ion conductor) lithium ion conducting glass ceramics (LICGC), $Li_{1+x+y}Al_xTi_{2-x}Si_yP_{3-y}O_{12}$, whose conductivity can reach the range of $1 \times 10^{-4} \, S \, cm^{-1}$. By melt-quenching method, the membrane can be easily fabricated and molded into different shapes on a large scale. Another morphological method is addition of impurities into solid conductors. As discussed above, by dispersing Al_2O_3 particles into the LiI matrix, which acts as a second phase, the Li ion conductivity can be largely enhanced as a result of space charge layer and percolation effect. Common types of space charge effects in composite materials are summarized in detail in Ref. [1], exhibiting the various kinds of interphase that are formed in different composite materials. A suitable combination of composites must be selected carefully when this system is applied.

References

[1] J. Gao, Y.-S. Zhao, S.-Q. Shi, H. Li: Lithium-ion transport in inorganic solid state electrolyte, *Chin. Phys. B* 25 (2016) 018211.

[2] J. C. Bachman, S, Muy, A. Grimaud, H.-H. Chang, N. Pour, S. F. Lux, O. Paschos, F. Maglia, S. Lupart, P. Lamp, L. Giordano, S-H, Yang: Inorganic solid-state electrolytes for lithium batteries: Mechanisms and properties governing ion conduction, *Chem. Rev.* 116 (2016) 140–162.

[3] K. Xu: Nonaqueous liquid electrolytes for lithium-based rechargeable batteries, *Chem. Rev.* 104 (2004) 4303–4418.

[4] J. T. Kummer: β-alumina electrolytes, *Prog. Solid State Chem.* 7 (1972) 141–175.

[5] R. Tärneberg, A. Lundn: Ion diffusion in the high temperature phases Li_2SO_4, $LiNaSO_4$, $LiAgSO_4$ and $Li_4Zn(SO_4)_3$, *Solid State Ionics* 90 (1996) 209–220.

[6] F. A. Kroeger: *The Chemistry of Imperfect Crystals*, 2nd Edn. (North-Holland, Amsterdam, The Netherlands, 1974).

[7] K. Sasaki: Influence of defect structure on the thermal conduction characteristic of YSZ polycrystalline, *Netsu Sokutei* 40 (2013) 65–70.

[8] W. Jost: Diffusion and electrolytic conduction in crystals, *J. Chem. Phys.* 1 (1933) 466–475.

[9] N. F. Mott, M. J. Littleton: Conduction in polar crystals. I. Electrolytic conduction in solid salts, *Trans. Faraday Soc.* 34 (1936) 485–499.

[10] M. Kotobuki, M. Koishi: Preparation of $Li_{1.5}Al_{0.5}Ti_{1.5}(PO_4)_3$ solid electrolyte via a sol–gel route using various Al sources, *Ceramics Intl.* 39 (2013) 4645–4649.

[11] J. Maier: Ionic conduction in space charge regions, *Prog. Solid State Chem.* 23 (1995) 171–263.

[12] H. L. Tuller: Ionic conduction in nanocrystalline materials, *Solid State Ionics* 131 (2000) 143–157.

[13] S. Kim, J. Maier: On the conductivity mechanism of nanocrystalline ceria, *J. Electrochem. Soc.* 149 (2002) J73–J83.

[14] J. Maier: Defect chemistry and ionic conductivity in thin films, *Solid State Ionics* 23 (1987) 59–67.

[15] P. Lupetin, G. Gregori, J. Maier: Mesoscopic charge carriers chemistry in nanocrystalline $SrTiO_3$, *Angew. Chem. Int. Ed.* 49 (2010) 10123–10126.

[16] C. Wagner: The electrical conductivity of semiconductors involving inclusions of another phase, *J. Phys. Chem. Solids* 33 (1972) 1051–1059.

[17] J. Maier: Space charge regions in solid two-phase systems and their conduction contribution. I. Conductance enhancement in the system ionic

conductor 'inert' phase and application on $AgCl:Al_2O_3$ and $AgCl:SiO_2$, *J. Phys. Chem. Solids* 46 (1985) 309–320.

[18] U. Lauer, J. Maier: Ionic contact equilibria in solids-implications for batteries and sensors, *Ber. Bunsen-Ges. Phys. Chem.* 94 (1990) 973–978.

[19] M. Bendahan, C. Jacolin, P. Lauque, J.-L. Seguin, P. Knauth: Morphology, Electrical conductivity, and reactivity of mixed conductor CuBr films: Development of a new ammonia gas detector, *J. Phys. Chem. B* 105 (2001) 8327–8333.

[20] J. S. Thokchom, B. Kumar: The effects of crystallization parameters on the ionic conductivity of a lithium aluminum germanium phosphate glass-ceramic, *J. Power Sources* 195 (2010) 2870–2876.

[21] K. Takada, N. Ohta, L. Q. Zhang, K. Fukuda, I. Sakaguchi, R. Z. Ma, M. Osada, T. Sasaki: Interfacial modification for high-power solid-state lithium batteries, *Solid State Ionics* 179 (2008) 1333–1337.

[22] X. Xu, K. Takada, K. Watanabe, I. Sakaguchi, K. Akatsuka, B. T. Hang, T. Ohnishi, T. Sasaki: Self-organized core-shell structure for high-power electrode in solid-state lithium batteries, *Chem. Mater.* 23 (2011) 3798–3804.

[23] N. Ohta, K. Takada, I. Sakaguchi, L. Q. Zhang, R. Z. Ma, K. Fukuda, M. Osada, T. Sasaki: $LiNbO_3$-coated $LiCoO_2$ as cathode material for all solid-state lithium secondary batteries, *Electrochem. Comm.* 9 (2007) 1486–1490.

[24] Y. Seino, T. Ota, K. J. Takada: High rate capabilities of all-solid-state lithium secondary batteries using $Li_4Ti_5O_{12}$-coated $LiNi_{0.8}Co_{0.15}Al_{0.05}O_2$ and a sulfide-based solid electrolyte, *J. Power Sources* 196 (2011) 6488–6492.

[25] J. Haruyama, K. Sodeyama, L. Han, K. Takada, Y. Tateyama: Space-charge layer effect at interface between oxide cathode and sulfide electrolyte in all-solid-state lithium-ion battery *Chem. Mater.* 26(14) 4248–4255

[26] A. J. Bhattacharyya, J. Maier: Second phase effects on the conductivity of non-aqueous salt solutions: "Soggy sand electrolytes", *Adv. Mater.* 16 (2004) 811–814.

[27] C. C. Liang: Conduction characteristics of the lithium iodide-aluminum oxide solid electrolytes, *J. Electrochem. Soc.* 120 (1973) 1289–1292.

Chapter 5

Crystalline Li-ion Conductive Ceramics

The Crystalline electrolytes possess a rigid skeleton composed of complex anion polyhedra that are long-range ordered. Li-ion can migrate in the skeleton with low activation energy. Li-ion conductivity of crystalline electrolytes is generally 1 to 2 orders higher than that of amorphous/glass electrolytes. However, preparation of a single-crystal solid electrolyte is costly and consumes a lot of energy and time. Therefore, polycrystalline solid electrolytes have usually been used. In the polycrystalline electrolytes, grain-boundary conductivity is generally lower than that in the crystal (bulk conductivity), particularly for Li_3N, β-Al_2O_3 and perovskite compounds which have 2D anisotropic conduction path. On the other hand, 3D ionic pathways can theoretically avoid the conductivity degradation at the grain boundaries in polycrystalline ceramics. Many polycrystalline Li-ion conductors have been researched so far. Among them, oxide and sulfide Li-ion conductors have been highly anticipated to serve as solid electrolytes. In this chapter, crystalline oxide and sulfide Li-ion conductors are reviewed.

5.1 Oxide Li-ion Conductive Ceramics

Compared with other inorganic solid Li-ion conductors such as halides and sulfides, oxide Li-ion conductors are more stable in air and at high temperature. Thus, they can be easily used for

Table 5.1 Li-ion conductivity of crystal oxide solid electrolytes at room temperature.

Electrolyte	Conductivity ($S\,cm^{-1}$)	Structure	Reference
$Li_{1.3}Al_{0.3}Ti_{1.7}(PO_4)_3$	7×10^{-4}	NASICON	[1]
$Li_{1.5}Al_{0.5}Ge_{1.5}(PO_4)_3$	1.8×10^{-4}	NASICON	[2]
$Li_{0.34}La_{0.51}TiO_{2.94}$	1.4×10^{-3}	Perovskite	[3]
$Li_{0.35}La_{0.55}TiO_3$	1.1×10^{-3}	Perovskite	[4]
$Li_5La_3Ta_2O_{12}$	3.4×10^{-6}	Garnet	[5]
$Li_7La_3Zr_2O_{12}$	4.67×10^{-4}	Garnet	[6]
Al-added $Li_7La_3Zr_2O_{12}$	1.4×10^{-4}	Garnet	[7]
Ge-doped $Li_7La_3Zr_2O_{12}$	8.3×10^{-4}	Garnet	[8]
$Li_{7-x}La_3Zr_{1.5}Ta_{0.5}O_{12-d}$	1.3×10^{-3}	Garnet	[9]
Tetragonal $Li_7La_3Zr_2O_{12}$	1.6×10^{-6}	Garnet	[10]
$Li_{5.25}La_3Ta_{1.75}Ge_{0.25}O_{12}$	8.4×10^{-5}	Garnet	[11]

manufacturing. Moreover, raw materials for the oxide Li-ion conductors are also readily available. Among the oxide Li-ion conductors, NASICON (Na super ion conductor), perovskite and garnet types have been extensively studied due to their high Li-ion conductivity. They are normally prepared through calcination of a mixture of raw materials to obtain electrolyte powder. Then, the powder is pelletized and sintered at high temperature to reduce grain-boundary resistance. Table 5.1 shows room temperature conductivity of typical crystalline oxide solid electrolytes. The conductivities of most oxide solid electrolytes in the table lie at about $1 \times 10^{-3} - 10^{-4}\,S\,cm^{-1}$, which is about one order lower than that of sulfide Li-ion conductors described later. Despite the inferior conductivity, the oxide Li-ion conductors have gained much attention for their practical use due to their stability in air, ease of handling and low production cost.

5.1.1 *NASICON-type solid electrolyte*

NASICON-type crystallographic structure, known as sodium superionic conductor, was first identified as early as the 1960s. In 1976, Goodenough *et al.* reported the pioneering work on the fast Na^+-ion transport system of $Na_{1+x}Zr_2P_{3-x}Si_xO_{12}$ [12]. The total ionic conductivity was measured as high as $10^{-3}\,S\,cm^{-1}$, identifying

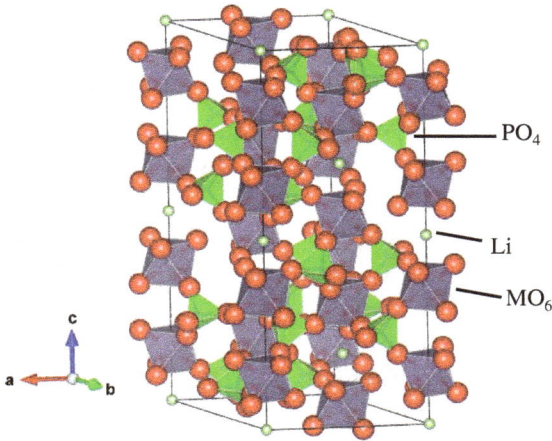

Fig. 5.1   NASICON structure of $LiM_2(PO_4)_3$.

that this framework structure could be a possible candidate for fast Li-ion transport by simply replacing the Na ion with a Li-ion. Since then, extensive efforts have been made to investigate the Li analogs $LiM_2(PO_4)_3$ (M = Ti, Ge, Zr, Hf, Sn, etc.) with varying compositions.

Figure 5.1 depicts the typical NASICON structure of Li-ion conductors which are depicted with the general formula of $LiM_2(PO_4)_3$ (M:tetravalent ion) with a space group of rhombohedral phase [13]. As can be seen, there are two types of polyhedra, MO_6 octahedra and PO_4 tetrahedra, in this structure. These polyhedral are linked to each other at their corners and form the $[M_2(PO_4)_3]^-$ rigid skeleton which is stabilized by electrons from Li-ions. Li-ion migrates through the tunnel three dimensionally in the structure. Mobile Li-ions are distributed to two sites, M' and M'' sites. M' sites are distributed along the c-axis between two $[MO_6]$ octahedra with a distorted octahedral co-ordination. M'' sites are located between $[PO_4]$ teterahedra and are connected to $[MO_6]$ octahedral by prismatic trigonal co-ordination [14]. The M' and M'' sites are connected together and form a 3D pathway for Li-ion conduction. The Li-ion must struggle going through a bottleneck when it jumps between adjacent sites. The bottleneck size depends greatly on the type, character and size of the skeleton ions [15] and is influenced by the distribution and concentration of mobile ions at the sites (M' and M'') [2]. The NASICON with abundant variety of transition metals, for example, M = Ti, Zr, Ge Sn and Hf, has been tested for studying the relationship among chemical composition, crystal structure and ionic conductivity [16]. Among the various types of Li-ion conductive NASCION solid electrolytes, in particular, $LiZr_2(PO_4)_3$, $LiTi_2(PO_4)_3$ and $LiGe_2(PO_4)_3$ have been thoroughly investigated.

$LiZr_2(PO_4)_3$ was expected to exhibit high Li-ion conductivity since its mother framework $[Zr_2(PO_4)_3]^-$ is same as Na-ion conductive NASICON [17]. However, $LiZr_2(PO_4)_3$ is stable in triclinic phase at room temperature, which does not support fast Li-ion conduction, due to smaller size of Li^+ than Na^+ [18]. In fact, the triclinic $LiZr_2(PO_4)_3$ showed low Li-ion conductivity of $\sim 10^{-8} \, S \, cm^{-1}$ [19, 20]. This value was much lower than the sodium analogue ($\sim 10^{-3} \, S \, cm^{-1}$) [12]. High conductive rhombohedral phase of $LiZr_2(PO_4)_3$ appears only above 50°C [21]. It was reported that $LiZr_2(PO_4)_3$ prepared at higher than 1100°C adopts high conductive rhombohedral phase but would undergo a phase transition to triclinic phase when cooling down to below 55°C [18, 19]. This result is in accordance with Arbi *et al.*'s work [20]. By neutral diffraction techniques, they also proved the phase transition phenomenon between the triclinic C-1 phase and the rhombohedral R-3c phase. Both phases exist in both heating and cooling processes at about 37°C. To stabilize the favorable rhombohedral phase at room temperature, Xie *et al.* doped $LiZr_2(PO_4)_3$ by substituting 5% Zr^{4+} with Ca^{2+}; the introduction of Ca^{2+} stabilizes the high conductive rhombohedral phase at room temperature, thereby giving a bulk conductivity of $1.2 \times 10^{-3} \, S \, cm^{-1}$, which is comparable to that of $Li_{1.3}Ti_{1.7}Al_{0.3}(PO_4)_3$ that is being expected to function as a solid electrolyte for novel types of cell configurations [22]. Moreover, Li *et al.* reported that the rhombohedral NASICON phase of $LiZr_2(PO_4)_3$ can also be stabilized by the introduction of Y^{3+}. The dopant Y^{3+} modified the sizes of M' and M'' to facilitate the Li^+ ion transport in the framework. The bulk and total

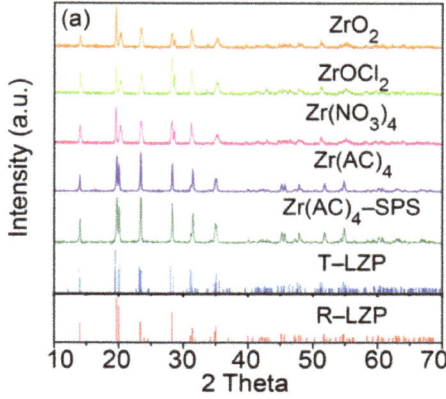

Fig. 5.2 XRD patterns of $LiZr_2(PO_4)_3$ prepared by using different zirconium sources.

conductivities of $Li_{1.15}Y_{0.15}Zr_{1.85}(PO_4)_3$ sintered by spark plasma sintering (SPS) technique were 1.4×10^{-4} and $0.71 \times 10^{-4}\,S\,cm^{-1}$ at 25°C, respectively [23]. More recently, encouraging progress was made by the same group. They claimed that instead of applying Y^{3+} and Ca^{2+} doping strategy, a pure $LiZr_2(PO_4)_3$ with high conductive rhombohedral phase can be obtained by using acetate as a zirconium source, which supplies particles with the rhombohedral phase in a single firing at 900°C, while in other zirconium sources a triclinic C-1 phase was obtained (Fig. 5.2). Through this simpler route, the product gives a bulk Li-ion conductivity of $2 \times 10^{-4}\,S\,cm^{-1}$ at 25°C [24]. These results are promising, but further work is required for $LiZr_2(PO_4)_3$ solid electrolyte to realize its applications in solid state Li battery. In addition to the $LiZr_2(PO_4)_3$ system, other systems, $LiHf_2(PO_4)_3$ and $LiSn_2(PO_4)_3$, are also crystallized in the low conductive triclinic lattice at room temperature.

The systems with $M = Ti^{4+}$(LTP) which exhibited rhombohedral lattice and relatively high Li-ion conductivity (about $10^{-5}\,S\,cm^{-1}$ at room temperature) have been most widely investigated [25]. This could be explained by matching the bottleneck size composed of TiO_6 octahedra with the size of the Li-ion. It was reported that the Ti^{4+}-based skeleton with a cell volume of 1310 Å provides the best fit for Li-ion transportation. Subramanian *et al.* have investigated the conductivities of $Li_{1+x}Zr_2(PO_4)_3$, $Li_{1+x}Zr_{2-x}Ti_x(PO_4)_3$ ($x = 0.1$–2), $LiTi_2(PO_4)_3$, $Li_{1+x}Ti_{2-x}Sc_x(PO_4)_3$ and $Li_{1+x}Hf_{2-x}In_x(PO_4)_3$ and reported that substitution of M^{3+} for M^{4+} in the series of $Li_{1+x}M^{3+}M^{4+}(PO_4)_3$ (M^{4+} = Zr, Ti and Hf) largely enhanced the Li-ion conductivity [26]. Since then, many dopants and substituents have been examined, e.g. Mg, In, Ga, Sc, Al, La, Y and Sn, for $M^{4+} = Ti^{4+}$ [1, 26–31], Nb, Ta, Y and In for $M^{4+} = Zr^{4+}$ [32], Al, Ga, Sc and In for $M^{4+} = Ge^{4+}$ [33,34] and In and Sc for $M^{4+} = Hf^{4+}$ [35–37]. However, the reason for conductivity enhancement by the M^{3+} introduction is still on debate. Aono *et al.* studied conductivity of $Li_{1+x}M_xTi_{2-x}(PO_4)_3$ and reported the increase of conductivity by M^{3+} introduction (Fig. 5.3) [1]. They claimed this increase was due mainly to decreased porosity and increase of grain-boundary conductivity [1,34,36,38]. Also, they reported the activation energy for bulk conductivity was essentially the same independent of M^{3+} element. This is also supported by Wang's work on Al-LTP (LATP) [39]. On the contrary, it has been proposed that extra Li-ions by charge compensation in M^{3+} substitution tuned Li-ion conduction pathway,

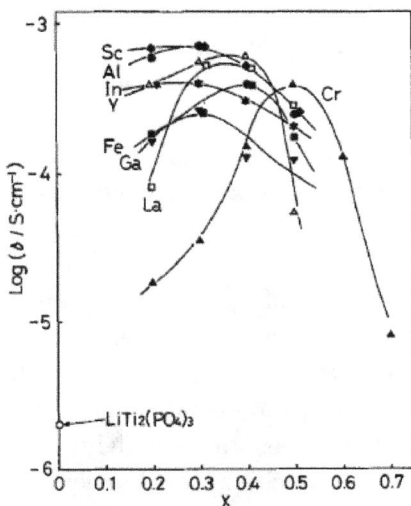

Fig. 5.3   Li-ion conductivity of $Li_{1+x}M_xTi_{2-x}(PO_4)_3$ system.

resulting in enhancement of bulk conductivity [40,41]. Chowdari *et al.* described that the difference of bottleneck size between $LiGe_2(PO_4)_3$ and $Li_{1+x}Al_xGe_{2-x}(PO_4)_3$was small since the ionic radii of Ge^{4+} and Al^{3+} are almost equal (0.530 and 0.535 Å, respectively) [42]. Therefore, they claimed increase in Li-ion concentration would cause conductivity enhancement. Among the LATP with various amounts of Al-doping, the highest conductivity of $3 \times 10^{-3}\,S\,cm^{-1}$ at room temperature was obtained in $Li_{1.3}Al_{0.3}Ti_{1.7}(PO_4)_3$ [30]. In addition to element substitution, ionic conductivity of LTP-based systems could also be improved considerably by appropriate sintering conditions and synthesis routes. Aono *et al.* investigated the influence of addition of various lithium salts during sintering, e.g. Li_3BO_3, Li_3PO_4, $LiNO_3$, $LiCl$ and Li_2SO_4. They found that the addition of lithium salts contributed to the densification of sintered pellets, leading to the enhancement of the ionic conductivity [43, 44]. Preparation techniques are also critical to obtain high conductive LTP system. Compared with conventional solid state reaction, sol–gel and melt-quenching process have been proven to be a more effective way to enhance the total ionic conductivity. The ionic conductivity of $Li_{1+x}Al_xTi_{2-x}(PO_4)_3$ glass ceramic prepared by the melting quench method is greatly improved in comparison with conventional synthesis route; a maximum conductivity of $1.3 \times 10^{-3}\,S\,cm^{-1}$ can be achieved. Zhang *et al.* also reported that a high conductivity of $1.22 \times 10^{-3}\,S\,cm^{-1}$ was achieved for an impurity phase-free $Li_{1.4}Al_{0.4}Ti_{1.6}(PO_4)_3$ using sol–gel method [45]. Moreover, $Li_{1+x}Al_xTi_{2-x}(PO_4)_3$ solid electrolytes were also reported to be prepared by co-precipitation method [46, 47], conventional sintering assisted with nanosized precursor powders synthesized by sol–gel method, spark plasma sintering (SPS) [48], etc. These results suggest that the main factors determining the ionic conductivity are the grain-boundary conductivity and relative density of bulk materials. By using SPS technique, maximum conductivities at room temperature of $1.39 \times 10^{-3}\,S\,cm^{-1}$ and $1.12 \times 10^{-3}\,S\,cm^{-1}$ for grain-boundary conductivity and total conductivity, respectively, were obtained. This high total conductivity is attributed to the compact pellets by minimizing the porosity and grain boundary [49].

Although the LTP-based oxides are attractive for use as solid electrolytes owing to their high total ionic conductivity, the facile reduction of Ti^{4+} ions at potentials below $2.5\,V$ vs. Li^+/Li would restrict their application [50,51]. When the LTP-based oxides contact with low-potential anodes like Li metal or C_6Li, the reduction of Ti^{4+} ions to Ti^{3+} induces electronic conduction and leads to a short circuit in batteries. The ionic radius of Ge^{4+} is similar to Ti^{4+} and Ge^{4+} is considered insensitive to reduction compared with Ti^{4+}. Therefore, Ge-based NASICON ($LiGe_2(PO_4)_3$) also has been intensively studied. Especially, $Li_{1+x}Al_xGe_{2-x}(PO_4)_3$ (LAGP) is expected to serve for battery applications due to the high Li-ion conductivity ($\sim 10^{-4}\,S\,cm^{-1}$) [2]. The rhombohedral LAGP can be obtained in the range $0 \le x \le 0.6$ [33], and the highest Li-ion conductivity was obtained at $x = 0.5$. LAGP had been considered to be in stable contact with Li metal [52, 53]. However, instability of LAGP under reduction potential below 0.648–$0.85\,V$ vs. Li^+/Li has been found in recent research [45, 54]. $LiZr_2(PO_4)_3$(LZP) is expected to be electrochemically stable in contact with Li metal. LZP has been paid attention to again recently with regard to this point. Li *et al.* recently found that a room-temperature LZP with rhombohedral phase will form a Li^+-ion conducting solid electrolyte interphase when in contact with lithium. The thin layer contains Li_8ZrO_6 and Li_3P that is wet by the lithium anode and also wets the LZP electrolyte itself. This causes small interfacial resistance for Li-ion transfer and inhibits Li dendrite formation during cycling [24]. However, as mentioned above, the formation of low ion conductive triclinic phase at room temperature must be overcome to use this material. Further study is needed to realize its possible applications. It was reported that the high conductive rhombohedral phase could be stabilized by substitution of Zr^{4+} with Ca^{2+} and Y^{3+} at room temperature [23, 55].

NASICON-type Li-ion conductors with high stability in air, high ionic conductivity and wide electrochemical window have been extensively investigated in the past decades. However, some unfavorable properties still limit its potential for wide applications. In NASICON-type compounds, a lot of possible combination of substituent and host elements can be considered. New NASICON compounds with high Li-ion conductivity and those that are stable in contact with Li metal could be developed on the basis of the above-mentioned researches. The Li-ion conductivity of representative NASICON-type Li-ion conductor is shown in Table 5.2.

Table 5.2 Li-ion Conductivities of Representative NASION-type Li-ion Conductor.

Formula	Synthesis route	Ionic conductivity $(S\,cm^{-1})$	Activation energy (eV)	Ref.
$LiTi_2(PO_4)_3$	Solid-state reaction	7.9×10^{-8}		[27]
$Li_{1.3}Al_{0.3}Ti_{1.7}(PO_4)_3$	Solid-state reaction	7×10^{-4}		[1]
	Melting-quench	1.3×10^{-3}		[56]
	Solution chemistry	1.6×10^{-4}		[48]
$Li_{1.4}Al_{0.4}Ti_{1.6}(PO_4)_3$	Melting-quench	3.8×10^{-5}	0.52	[52]
	Sol–gel	6.13×10^{-4}	0.29	[57]
	SPS sintering	1.12×10^{-3}	0.25	[58]
$Li_{1.5}Al_{0.5}Ti_{1.5}(PO_4)_3$	Co-precipitation	1.5×10^{-4}		[47]
$LiZr_2(PO_4)_3$	Solid-state reaction	$< 10^{-9}$		[27]
$Li_{1.2}Zr_{1.9}Ca_{0.1}(PO_4)_3$	Solid-state reaction	1.2×10^{-4}	0.48	[22]
$Li_{1.15}Y_{0.15}Zr_{1.85}(PO_4)_3$	SPS sintering	7.1×10^{-5}	0.39	[23]
$LiGe_2(PO_4)_3$	Solid-state reaction	$2 \times 10^{-8}(60°C)$	0.72	[42]
$Li_{1.5}Al_{0.5}Ge_{1.5}\ (PO_4)_3$	Solid-state reaction	3.5×10^{-5}	0.33	[33]
	Solid-state reaction	2.4×10^{-4}		[34]
	Melting-quench	4×10^{-4}	0.36	[56]
	Melting-quench	5.08×10^{-3}	0.31	[59]
	Melting-quench	5.8×10^{-4}	0.28	[60]
	Melting-quench and SPS sintering	1.33×10^{-4}	0.38	[61]
	Melting-quench	5.21×10^{-4}	0.34	[62]
	Sol–gel	1.8×10^{-4}		[2]
$Li_{1.5}Al_{0.5}Ge_{1.5}$ $(PO_4)_3$-$0.05Li_2O$	Melting-quench	7.25×10^{-4}	0.31	[53]
$Li_{1.4}Al_{0.4}Ge_{1.6}(PO_4)_3$	Solid-state reaction	1.3×10^{-4}		[63]
	Sol–gel	1.22×10^{-3}		[45]
$LiHf_2(PO_4)_3$	Solid-state reaction	3.42×10^{-6}	0.77	[37]
$Li_{1.2}Fe_{0.2}Hf_{1.8}(PO_4)_3$	Solid-state reaction	1.7×10^{-4}	0.42	[36]

5.1.2　*Perovskite*

Commonly known for their dielectric and ferroelectric properties, perovskite type materials also comprise a family of solid Li-ion conductors. The ideal perovskite structure with a general formula of ABO_3 has a cubic symmetry unit cell with the space group Pm3m, where alkaline rare earth or earth metal ions occupy A sites at the corners of a cube, transitional metal ions occupy B sites at the center and the oxygen anions are at the face-center positions. In this structure, the A cation and the B cation are in 12-fold co-ordination and 6-fold co-ordination (BO_6), respectively, as shown in Fig. 5.4. The studies on solid Li-ion conductors with perovskite structure began with the investigations on Lithium Lanthanum Titanate (LLT). Lithium lanthanum titanate (LLT) (A = Li, La and B = Ti) with a general formula of $Li_{3x}La_{(2/3)-x}\square_{1/3-2x}TiO_3$ ($\square$: lattice vacancy) and its related materials are also known as fast Li-ion conductive ceramics. Brous *et al.* first reported the synthesis of cubic perovskite $Li_{0.5}La_{0.5}TiO_3$ by replacement of A-site in perovskite-type

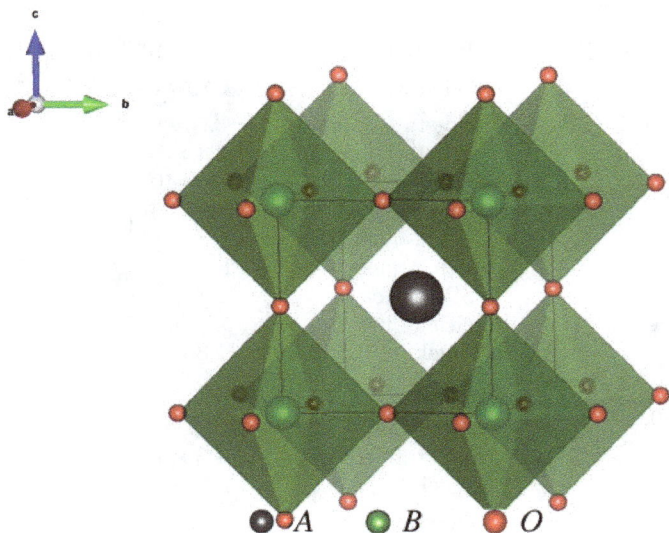

Fig. 5.4　Structure of ideal perovskite material.

alkaline-earth titanates ($ATiO_3$, A = Ca, Ba, Sr) with La^{3+} and Li^+ [64]. However, this was hexagonal phase confirmed later by the neutron diffraction [65]. The Li-ion motion of LLT was found by Inaguma *et al.* through their observation of increase in capacitance with temperature, large dielectric loss and dielectric relaxation [66]. Belous *et al.* reported that $Li_{3x}La_{(2/3)-x}\square_{1/3-2x}TiO_3$ showed a perovskite-like structure in the range of $0.04 < x < 0.17$, in which the larger La^{3+}-ion plays the role of stabilizing the perovskite-like structure and the smaller Li-ion contributes as a charge carrier [67]. The bulk Li-ion conductivity of LLT was reported to be *ca.* $1 \times 10^{-3}\,S\,cm^{-1}$ at room temperature [66]. Since then, LLT has attracted many research groups due to its possible potential application as a solid electrolyte for various electrochemical devices.

The general formula of LLT is $Li_{3x}La_{(2/3)-x}\square_{1/3-2x}TiO_3$ ($0 < x < 0.167$) [68–70]. There are four phases of LLT depending on x. The orthorhombic and hexagonal phases appeared at $x < 0.06$ [71] and $x = 0.167$ [65,71], respectively. In the range of $0.06 < x < 0.15$, either cubic or tetragonal phase or a mixture of both phases is observed. These compounds are A-site-deficient perovskites, and the vacancy sites are partially filled with various amounts of Li-ions [72, 73]. Details of each structure can be found in a review by Stramare *et al.* [68]. Figure 5.5 depicts the ideal structure of perovskite

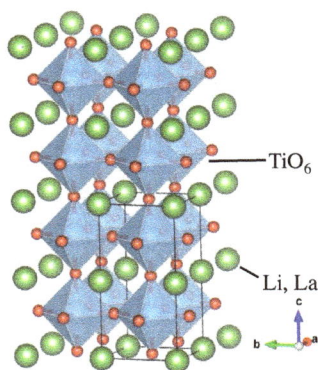

Fig. 5.5   Ideal structure of perovskite $Li_{3x}La_{(2/3)-x}\square_{1/3-2x}TiO_3$.

$Li_{3x}La_{(2/3)-x}\square_{1/3-2x}TiO_3$. The cubic and tetragonal phases are basically the same, and the difference is only in the distribution of Li and La in the A-site. The cubic phase (space group: Pm3m) with random distribution of Li and La can be obtained by quenching the sample above 1150°C [74]. The tetragonal phase (space group: P/4mmm) can be obtained by annealing the cubic phase below 1150°C, and it has alternative ordering of a La-rich layer and a La-poor layer along the c-axis [66, 75]. In this structure, the Li-ion migrates in La-poor layers due to the formation of a percolation pathway at A-sites for Li-ion, while the La-rich layers hinder the Li-ion migration, resulting in 2D Li-ion conduction in the LLT.

Although extensive fundamental studies have been carried out to demonstrate the high bulk ionic conductivity of LLT, the actual mechanism of ionic conduction is not yet clearly understood. It is widely accepted that the conduction behavior of LLT is thermally activated. However, it was observed that the Arrhenius plots of Li-ion conductivity for polycrystalline LLT showed a bending behavior at higher temperatures of around 400 K, as shown in Fig. 5.6. This discrepancy was attributed to phase transitions proposed by some researchers. They claimed that at higher temperature (400 K) the conduction process might involve two different activation energies, which can be explained by thermally assisted conduction mechanism.

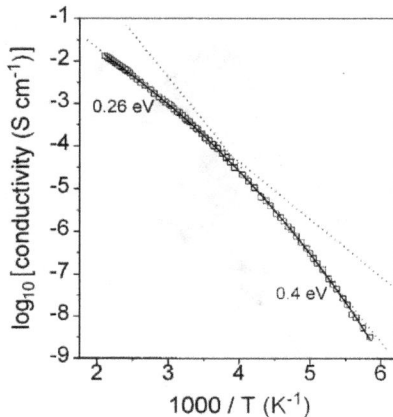

Fig. 5.6   Arrhenius plots of lithium ion conductivity of LLT by dc method [77].

According to this theory, conductivity data above 285 K could be fit by Vogel–Tammann–Fulcher relationship (VTF), which is related to the tilting of BO_6 octahedra, while at temperatures below 400 K, Arrhenius law is still followed [3, 67, 69, 76–78].

It has been found that the ionic conductivity is influenced by Li content [78]. The conductivity curve exhibited maximum value (*ca.* 1×10^{-3} S cm^{-1}) at around $x = 0.1$ (Li content is 0.3). the degree of ordering of cations and vacancies on the A sites also largely affects the ionic conductivity of LLT as well as the crystal structure. Harada *et al.* introduced an order parameter S that was defined for alternative arrangement in the tetragonal phase [79],

$$S = [R(\text{La-rich}) - R(\text{dis})]/[1 - R(\text{dis})].$$

Here, R(La-rich) and R(dis) mean occupancies of La^{3+} ions at the A site in the La-rich layers of the ordered form and in the (001) of disordered form, respectively. Detailed calculation procedures are described in Ref. [79]. The ionic conductivity decreased with increasing the order parameter S [76].

In addition, the substitution of La by other rare earth elements could also influence the conductivity. The ionic conductivity reduces with partial replacement of La by rare earth elements with smaller ionic radius, while it increases with partial substitution of La by Sr (larger ionic radius than that of La). Nonetheless, the influence of full or partial substitution of La on conductivity enhancement is limited. The number of vacancies and the bottleneck size of the skeleton have a major influence on the conduction behavior. In 2003, Morata-Orrantia *et al.* reported that high conductivities of 2.95×10^{-3} and 2.54×10^{-3} S/cm^{-1} were achieved for Al- and Sr-doped LLT, respectively, by tuning the pure LLT with optimal vacancy concentration first before doping [80]. Okumura *et al.* also improved the conductivity of LLT to 2.3×10^{-3} S/cm^{-1} by partially substituting F for O. This improvement was due to the expansion of bottleneck [81]. Other strategies, for example, creating highly disordered A-sites and minimizing lattice distortion, have also been proven to be effective ways for enhancement of bulk ionic conductivity [82].

Despite the high bulk ionic conductivity of LLT ($\sim 10^{-3}\,\text{S}\,\text{cm}^{-1}$), the grain-boundary conductivity of LLT is very low, normally less than $\sim 10^{-5}\,\text{S}\,\text{cm}^{-1}$ [75, 83, 84]. The total conductivity, which is composed of the bulk and grain-boundary conductivities, is largely influenced by the grain-boundary conductivity, and improvement of the grain-boundary conductivity is the key issue in the development of LLT. As mentioned above, Li-ions migrate two dimensionally in the La-poor layer of the LLT grain. The mismatch of the Li-ion conductive La-poor layer between randomly oriented neighboring grains makes a high potential barrier for Li-ion migration (Fig. 4.10), resulting in low grain-boundary conductivity [75]. The grain boundary in LLT was investigated in detail by using High-Angle Annular Dark Field Scanning Transmission Electron Microscope (HAADF-STEM) and electron energy loss spectroscopy [85]. The grain boundaries in LLT are largely different chemically and structurally from that inside the grains to compensate for the random orientation among the grains (Fig. 5.7). These chemical and structural differences from

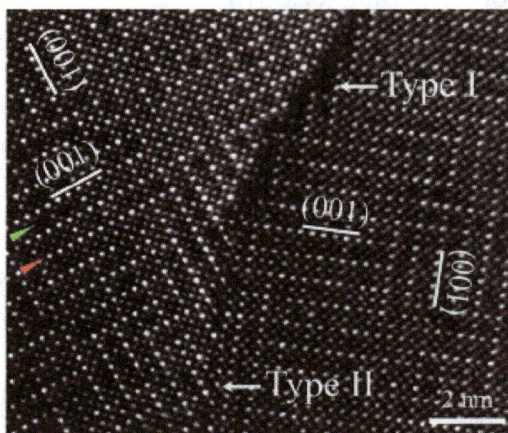

Fig. 5.7   HAADF-STEM image of grain boundary of LLT. La-poor layer and La-rich layers are shown by green and red arrows, respectively. The disconnectivity of ion conduction path can be seen clearly.

Source: Reproduced from Ref. [85] with permission from Royal Society of Chemistry.

perovskite network lead to the formation of a high energy barrier for Li-ion migration and cause low grain-boundary conductivity.

In order to improve the grain-boundary conductivity, two approaches have been attempted. One approach is improvement of grain-boundary conductivity by modification of the grain-boundary structure and composition. The grain-boundary conductivity of LLT sintered in oxygen was five times higher that sintered in air, which can be attributed to suppression of Li_2CO_3 formation caused by a reaction with CO_2 in air [86]. Also, LLT sintered with sacrificed powder, i.e. LLT pellet is covered by the sacrificed powder (normally same LLT powder as the pellet), exhibited higher grain-boundary conductivity than that without sacrificed powder due to restriction of Li volatilization at grain boundary [84]. Another approach is decrease of grain-boundary density by increase of sinterability of LLT pellet (increase of grain size) (Fig. 4.11). The resistance by the grain-boundary decreased with increase of sintering temperature due to increase in size of crystal grain (decrease in the number of grain boundaries). The total conductivities of LLT sintered at 1400°C and 1450°C were reported to be 2.4×10^{-4} and 5.0×10^{-4} S cm^{-1}, respectively [87]. Introduction of a heteroatom into the grain boundary also has been performed to improve the ionic conductivity. For example, formation of an amorphous silica layer at grain boundary improved the total conductivity to over 1×10^{-4} S cm^{-1}. The amorphous silica thin layer at the grain boundary not only works as a binder for the grains but also forms a 3D conduction region for Li-ion transport. This improvement of conductivity by amorphous silica layer was also observed in $Li_7La_3Zr_2O_{12}$ garnet solid electrolyte [88]. Other examples are described in detail by Stramare *et al* [68].

The intrinsic shortcoming of LLT is the ease of reduction of Ti^{4+} ions as mentioned in Section 5.1.1. The reduction of Ti^{4+} to Ti^{3+} in LLT occurs at potential of 1.8 V vs. Li$^+$/Li [89, 90]. This narrow electrochemical window restricts operation voltage of the all-solid-state battery using LLT as a solid electrolyte. The chemical substitution of LLT has been applied to widen the electrochemical window. Kuhn *et al.* observed suppression of Ti^{4+} reduction in LLT by Na substitution for the Li site [91]. Another problem encountered with LLT is the high Li losses during high temperature sintering that is required to obtain well-sintered LLT pellet, as this results in difficulties in controlling the Li-ion content and Li-ion conductivity.

5.1.3 *Garnet*

Garnet-type solid Li-ion conductors were derived from the ideal garnet structure with a general formula of $A_3B_2X_3O_{12}$, such as $Ca_3Al_2Si_3O_{12}$, where A sites are 8-fold co-ordinated, B sites are 6-fold co-ordinated and X sites are 4-fold co-ordinated. The first Li-containing garnet structure, $Li_3M_2Ln_3O_{12}$ (M = Te, W), was reported by Kasper *et al.* in 1969 [92]. The garnet structure is composed of a mixture of cations on filled square anti-prismatic, octahedral and tetrahedral sites in a 3:2:3 ratio like $Ca_3Al_2Si_3O_{12}$. The $Li_3Ln_3Te_2O_{12}$ has the same structure as this and showed low ionic conductivity of $\sim 10^{-5}\,\mathrm{S\,cm^{-1}}$ at 600°C [93]. In this structure, Li-ions exist only in the tetrahedral (24d) sites. Li-ions are tightly bound in the 24d sites (Fig. 5.8), resulting in low ionic conductivity However, Li-containing garnet has been known to form unusual stoichiometry with $Li_5Ln_3M_2O_{12}$ where M is a pentavalent cation such as Ta^{5+} [94], Nb^{5+} or Sb^{5+} [95,96]. (The garnets containing Li higher than three per formula unit are called Li-stuffed garnets).

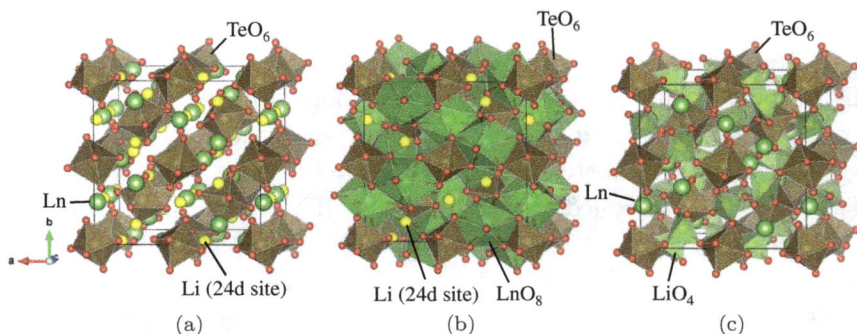

Fig. 5.8 Structure of Li-garnet where three structures are same. The difference is exhibition of polyhedral: (a) TeO_6 octahedra, (b) TeO_6 octahedra and LnO_8 dodecahedra, and (c) TeO_6 octahedra and LiO_4 tetrahedra.

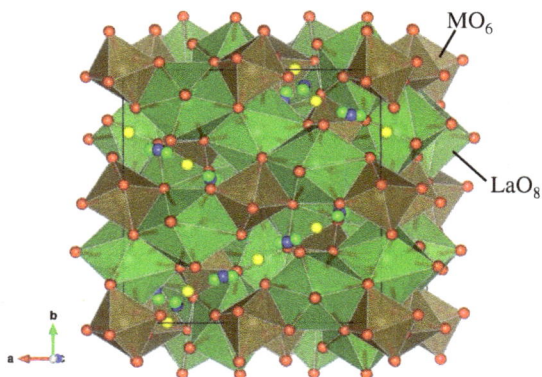

Fig. 5.9 Structure of $Li_5La_3M_2O_{12}$ (M = Nb, Ta), where yellow atoms show 24d (Li1) site, green represents 96h (Li2) sites and blue indicates 48g (Li3) sites.

In 2003 and also later, Thangadurai *et al.* has reported a series of garnet compounds, $Li_5La_3M_2O_{12}$, where M is a pentavalent cation, for the first time suggesting that garnet-type materials are potential candidates for Li-ion superionic conductors [5]. Among them, the Li-ion conductivity of the Li-stuffed garnets, such as $Li_5La_3Ta_2O_{12}$ and $Li_5La_3Nb_2O_{12}$, is around three orders of magnitude greater than that of $Li_3Nd_3Te_2O_{12}$ at room temperature, $\sim 10^{-5}\,S\,cm^{-1}$ [5, 97]. Figure 5.9 depicts the crystal structure of $Li_5La_3M_2O_{12}$. $Li_5La_3M_2O_{12}$ crystal belongs to the space group of Ia-3d in which excess Li-ions are in tetrahedral 24d site and distorted octahedral site (96 h/48 g). Both sites are partially filled [98]. The introduction of vacancies in the tetrahedral sites and partial occupation of Li-ion in the distorted octahedral sites where bonded to Li-ion loosely give rise to high Li-ion conductivity in the Li-stuffed garnets. M (= Nb and Ta) in $Li_5La_3M_2O_{12}$ can be replaced with Bi [99, 100], Sb [101, 102] and In [103]. In-substituted $Li_{5.5}La_3Nb_{1.75}In_{0.25}O_{12}$ exhibits a conductivity of $1.8 \times 10^{-4}\,S\,cm^{-1}$ at 50°C [103]. La also can be substituted by Pr, Nd [104], Ba [105] and K [103]. By partially substituting La with low valence ions, such as Ca, Sr and Ba [106, 107], the ionic conductivity of $Li_5La_3M_2O_{12}$ could be slightly improved. Ba-substituted $Li_6BaLa_2Ta_2O_{12}$ shows the highest conductivity, $4 \times 10^{-5}\,S\,cm^{-1}$, and the lowest activation energy,

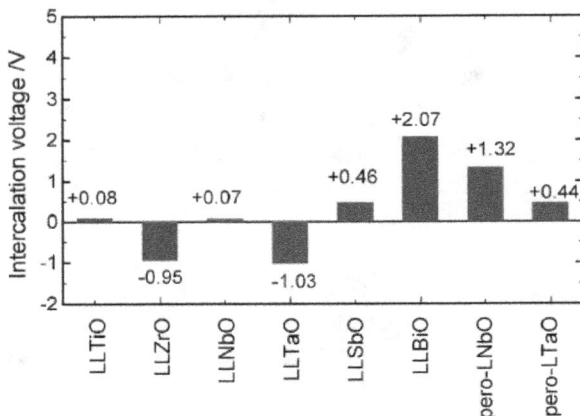

Fig. 5.10　Intercalation voltage of various solid electrolytes.

Source: Reproduced from Ref. [109] with permission from Royal Society of Chemistry.

0.40 eV, among these substituted garnets. The structural parameters of these $Li_5La_3M_2O_{12}$ garnets has been described in Ref. [108].

Among those garnets containing five Li atoms in the formula, $Li_5La_3Ta_2O_{12}$ (LLTa) is the most expected as a solid electrolyte due to stable in contact with Li metal proved by theoretically and experimentally [109]. Figure 5.10 shows the intercalation voltage of various Li-ion conductive ceramics calculated by the first-principles density functional theory. As mentioned in the previous section, LLTiO (LLT) perovskite has a positive intercalation voltage value, indicating that LLT is reduced by Li metal. However, LLTaO (LLTa) shows a negative value and is therefore expected to be stable in contact with Li metal. Figure 5.11 exhibits the observation of LLTa before and after contact with molten Li at about 200°C for 72 h [110]. No visible change was observed on contact with molten Li. This high stability of LLTa was also proved by XRD measurements. The XRD patterns after contact with molten Li for 72 h were consistent with that before contact, implying that structural change does not occur (Fig. 5.12) [110].

It is worth mentioning that perovskite $La_{1/3}TaO_3$ shows positive intercalation value by the first-principles density functional theory

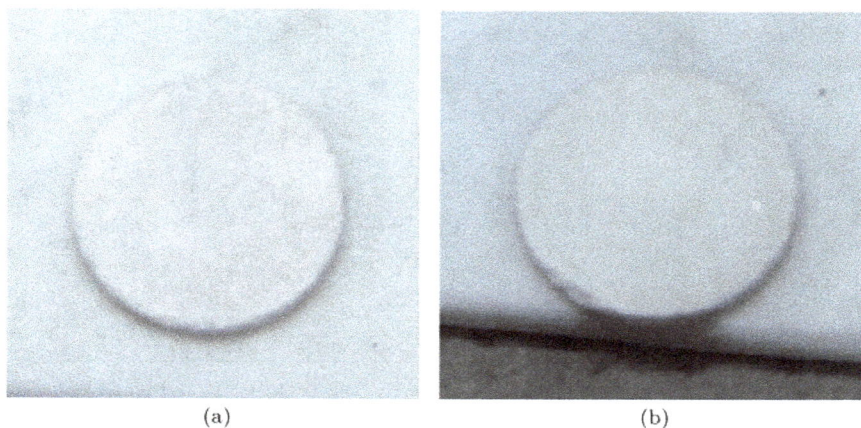

Fig. 5.11   Photos of LLTa pellet (a) before and (b) after contact with molten Li for 72 h.

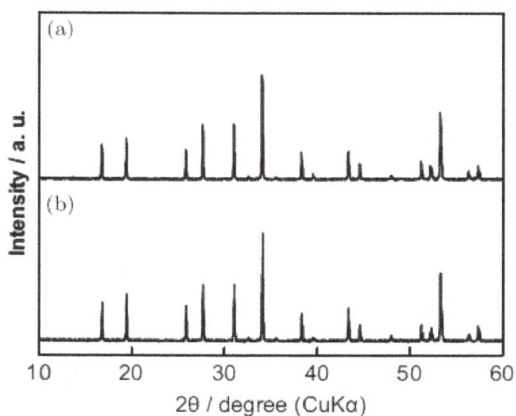

Fig. 5.12   XRD patterns of LLTa pellet (a) before and (b) after contact with molten Li for 72 h.

even though the perovskite $La_{1/3}TaO_3$ is composed of the same elements as LLTa. This behavior is well-matched with the experimental result. Figure 5.13 reveals photographs of garnet LLTa and perovskite $La_{1/3}TaO_3$ after they come into contact with molten Li. Color change was observed in the perovskite $La_{1/3}TaO_3$ within 3 min after the contact, while LLTa retained its white color. Therefore, it can be

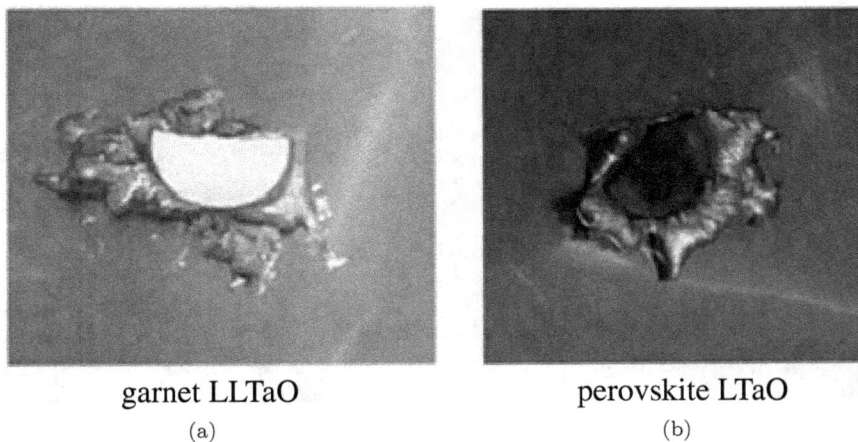

garnet LLTaO

(a)

perovskite LTaO

(b)

Fig. 5.13　Photographs of (a) garnet LLTa and (b) $La_{1/3}TaO_3$ after contact with molten Li.

Source: Reproduced from Ref. [109] with permission from Royal Society of Chemistry.

said that the garnet structure also contributes the excellent high stability of LLTa. This result gives us a hint that not only elements consisting of the ceramics but also their structure largely affect the properties of the ceramics in the development of the Li-ion conductive ceramics. Additionally, LLNbO ($Li_5La_3Nb_2O_{12}$) is not stable in contact with Li [5], and this is consistent with the first-principles density functional theory. It can be said that the first-principle density functional calculation is useful to predict the stability of the ceramics electrolyte. The electrochemical window of LLTa is also tested (Fig. 5.14). Redox peaks are not observed until 5 V vs. Li^+/Li except for Li deposition/dissolution peaks. Furthermore, LLTa does not react with $LiCoO_2$, which is a cathode material widely used in commercial Li batteries after calcination at 700°C for 2 h [110]. This high stability toward Li metal anode and $LiCoO_2$ cathode makes LLTa one of the promising ceramic electrolytes for all-solid-state battery. However, relatively low Li-ion conductivity has been a major concern for the application of LLTa for the all-solid-state batteries. Recently, Li-ion conductivity of LLTa has been

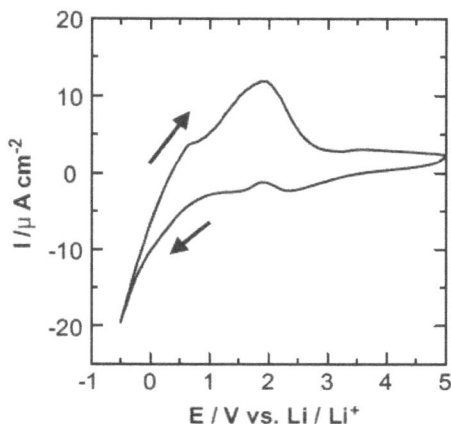

Fig. 5.14   Cyclic voltammogram of Li/LLTa/Au cell.

improvedby Ge-substitution for Ta site [11], which was confirmed by XRD patterns (Fig. 5.15). In the XRD patterns, no diffraction peak for Ge-containing materials was observed. In fact, the lattice parameter, Ge-substituted LLTa, was decreased with increase of Ge content due to smaller ion radius of Ge^{4+} (0.530 Å) than Ta^{5+} (0.640 Å).

By the substitution of pentavalent Ta^{5+} with tetravalent Ge^{4+}, additional Li can be introduced in the garnet lattice to maintain charge neutrality, resulting in increase of Li-ion concentration. As a consequence, Li-ion conductivity can be improved. The Li-ion conductivity of $Li_{5+x}La_3Ta_{2-x}Ge_xO_{12}$ is $8.4 \times 10^{-5}\,S\,cm^{-1}$, about one order higher than LLTa when $x = 0.25$. The Li-ion conductivity of LLTa is increased by Ge substitution, but electron conductivity is still negligible. Figure 5.16 shows the Hebb–Wagner polarization data of $Li_{5.25}La_3Ta_{1.75}Ge_{0.25}O_{12}$. The data was obtained by polarization of Li/solid electrolye/Au cell at 2.5 V. From the steady state current, the elecrton conductivity can be calculated. The electron conductivity was estimated to be $2.3 \times 10^{-7}\,S\,cm^{-1}$. This value was more than two orders of magnitude lower than that of the Li-ion condutivity. Therefore, the Li-ion transference number is considered to be unity.

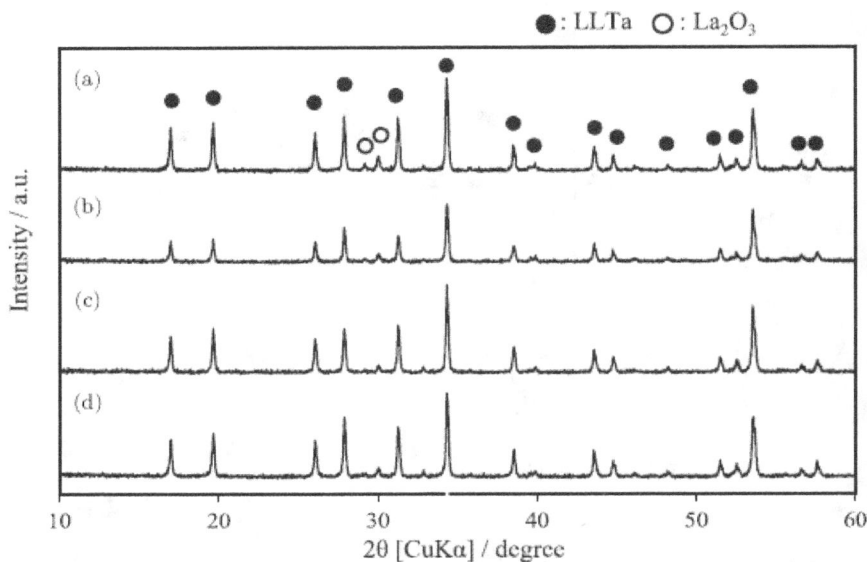

Fig. 5.15   XRD patterns of Ge-substituted LLTa sintered by SPS: (a) $x = 0.5$, 1100°C, (b) $x = 0.5$, 1050°C, (c) $x = 0.25$, 1100°C, (d) $x = 0.25$, 1050°C.

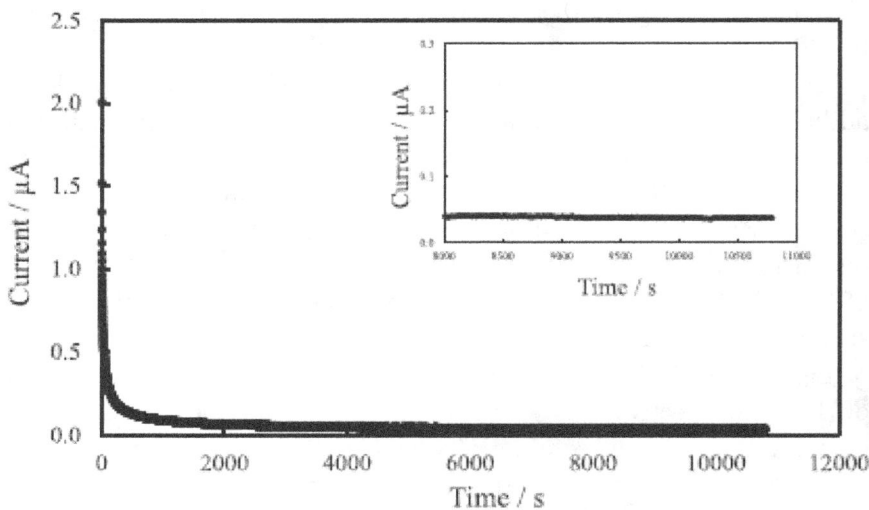

Fig. 5.16   Polarization curve of Li/LLTaGe/Au cell.

The significant progress of the garnet-related Li-ion conductors was achieved by the discovery of $Li_7La_3Zr_2O_{12}$(LLZ) in 2007 [6]. LLZ has a high bulk and total Li-ion conductivity of 4.67×10^{-4} and $2.44 \times 10^{-4}\,S\,cm^{-1}$, respectively, with low activation energy (*ca.* 0.3 eV).

This small difference between bulk and total conductivities implies the grain-boundary contribution to be comparable to the bulk one. Additionally, it must be mentioned that LLZ possesses advantages over other electrolytes such as good chemical and electrochemical stability against Li metal and wide electrochemical window [6].

LLZ was prepared by the solid state reaction from a mixture of Li_2CO_3, La_2O_3 and ZrO_2. Figure 5.17 shows the XRD patterns of the mixture calcined at various temperatures. Well-crystallized LLZ is obtained after calcination at 1230°C. The cyclic voltammogram of the Li/LLZ/Li cell clearly shows Li deposition and dissolution peaks, indicating that LLZ is stable in contact with Li metal and that reversible Li deposition/dissolution is possible (Fig. 5.18). Besides the solid state reaction, the sol–gel method has also been studied for LLZ preparation [111].

The structure of LLZ is shown in Fig. 5.19. The framework of LLZ garnet is composed of dodecahedral LaO_8 and octahedral ZrO_6.

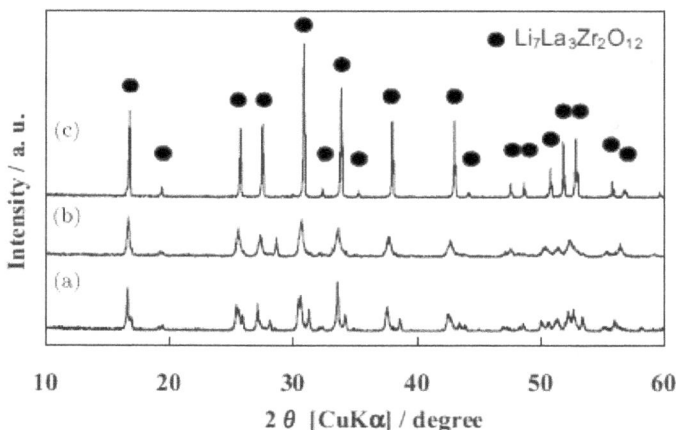

Fig. 5.17 XRD patterns of LLZ calcined at (a) 900°C, (b) 1125°C, and (c) 1230°C.

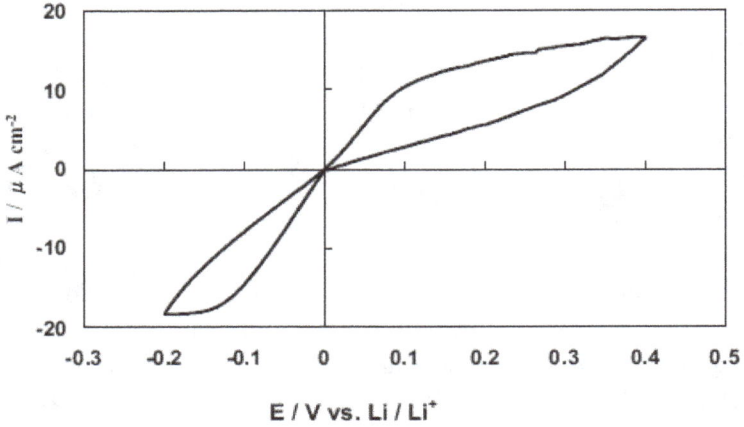

Fig. 5.18   Cyclic voltammogram of Li/LLZ/Li cell.

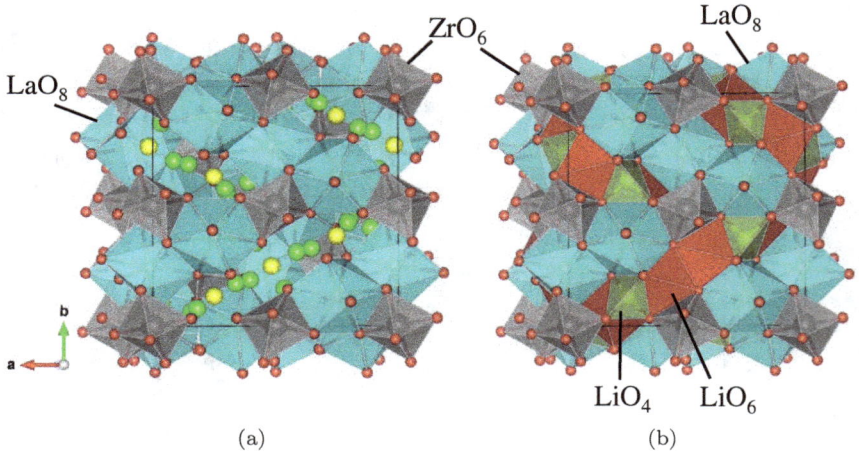

Fig. 5.19   Structure of LLZ.

In this structure, seven Li-ions are contained in a formula unit. When the Li-ions occupy both neighboring octahedral and tetrahedral sites, the Coulomb repulsion among the Li-ions displaces the Li-ion in the octahedral site from the 48 g position to the 96 h position [112] (Fig. 5.20). This displacement causes high ionic conductivity. The ionic conductivity increases with Li occupancy in 48 g/96 h sites [113].

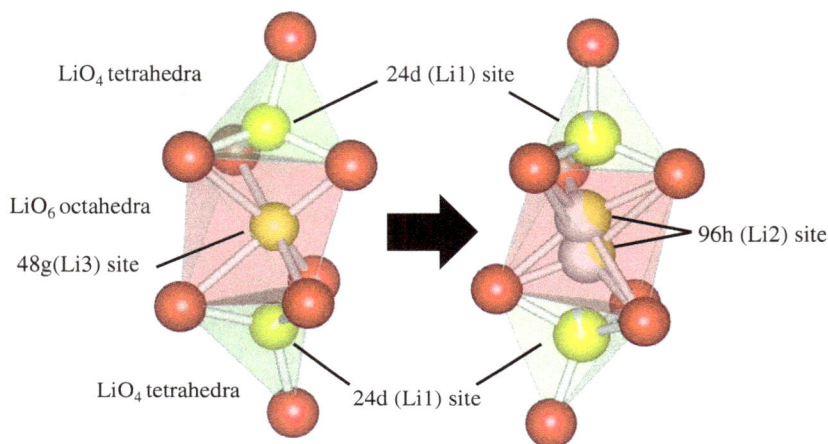

Fig. 5.20   48 g and 96 h positions.

Therefore, LLZ containing seven Li-ions in the formula unit exhibits about two orders higher ionic conductivity than LLTa with only five Li-ions in the formula unit.

Although LLZ is an attractive material for the solid electrolyte, tetragonal LLZ (space group $I4_1/acd$) which has low conductivity (Table 2.1) is thermodynamically more stable at room temperature [114] compared with the cubic phase (space group Ia-3d). Figure 5.21 shows the crystal structure of tetragonal LLZ (The structure of cubic LLZ is shown in Fig. 5.19). The garnet framework in the tetragonal LLZ is composed of $La(1)O_8$, $La(2)O_8$ and ZrO_6. Li-ions occupy three sites. Li(1) ions occupy tetrahedral 8a sites. Li(2) and Li(3) ions sit in octahedral 16f and 32 g sites, respectively [10]. The Li1, Li2 and Li3 sites are fully occupied. On the other hand, only two sites (tetrahedral 24d site and distorted octahedral 96 h sites) exist for Li-ions in the cubic phase as mentioned above. Li-ion arrangement of cubic and tetragonal LLZ is extracted and depicted in Fig. 5.22 [115]. In the cubic phase, the Li-ions construct a loop and only the Li1 site is shared by two loops as a junction between where the two Li2 sites exist. In the tetragonal phase as well, the Li1 site is shared by two loops; however, the loop arrangement is Li1–Li3–Li2–Li3–Li1sites (Fig. 5.22(b)). The distance between Li1 and Li2

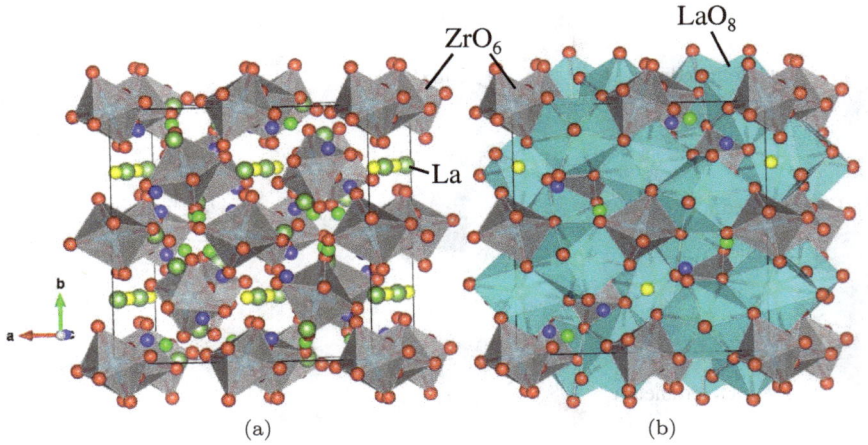

Fig. 5.21   Structure of tetragonal LLZ.

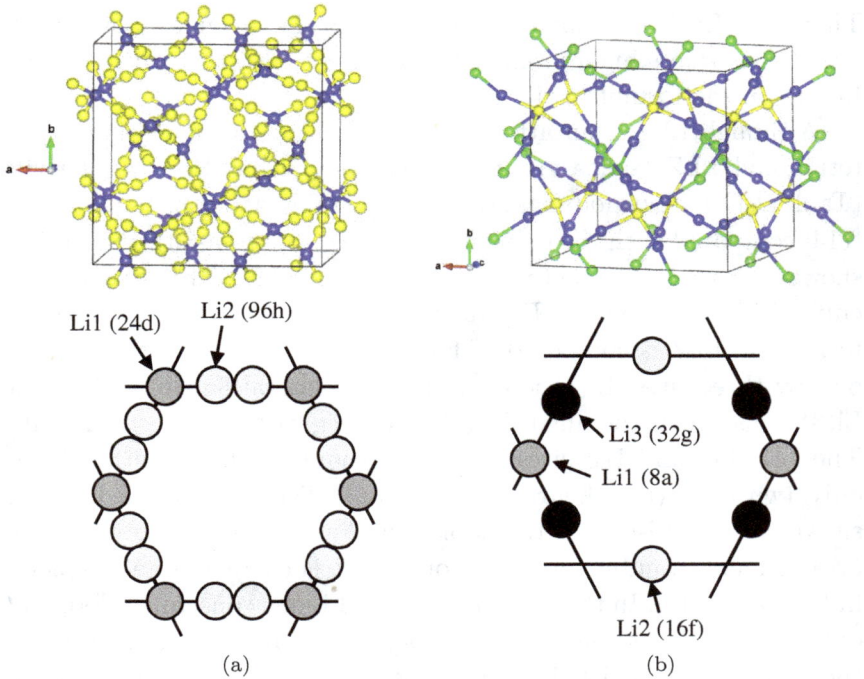

Fig. 5.22   Li-ion arrangement of (a) cubic and (b) tetragonal LLZ.

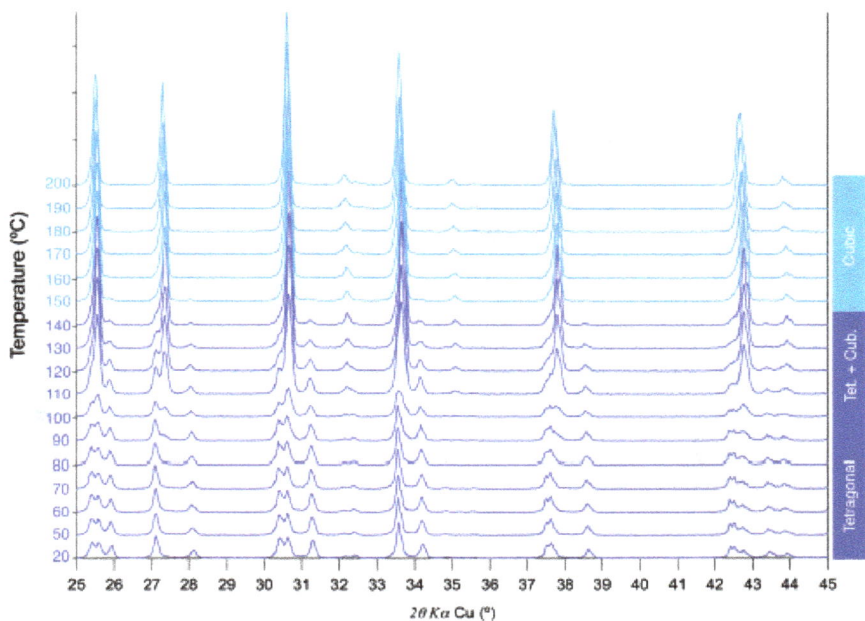

Fig. 5.23   HT-XRD patterns of tetragonal LLZ from 20°C to 200°C.

Source: Reproduced from Ref. [117] with permission from Royal Society of Chemistry.

site is 1.602(18) Å in the cubic phase. However, the Li–Li distance in the tetragonal phase is longer, >2.5 Å. This short distance in the cubic phase gives rise to the Coulomb repulsion and improves Li-ion mobility, resulting in the high ionic conductivity. Recently, the third phase of LLZ, which is called low temperature cubic (LT-cubic) phase, was found [116]. Figure 5.23 reveals the high-temperature XRD (HT-XRD) patterns of LLZ. The LT-cubic phase appeared when the tetragonal LLZ was heated at around 100°C in air [117]. Li_2CO_3 formation due to absorption of CO_2 in air by LLZ is suggested as a reason for the formation of LT-cubic phase [118], but there are still unknown points that preclude the explanation of the formation of LT-cubic phase. Further research is required regarding this. Upon further heating, the LT-cubic phase disappeared and the tetragonal phase appeared instead at around 450°C (Fig. 5.24).

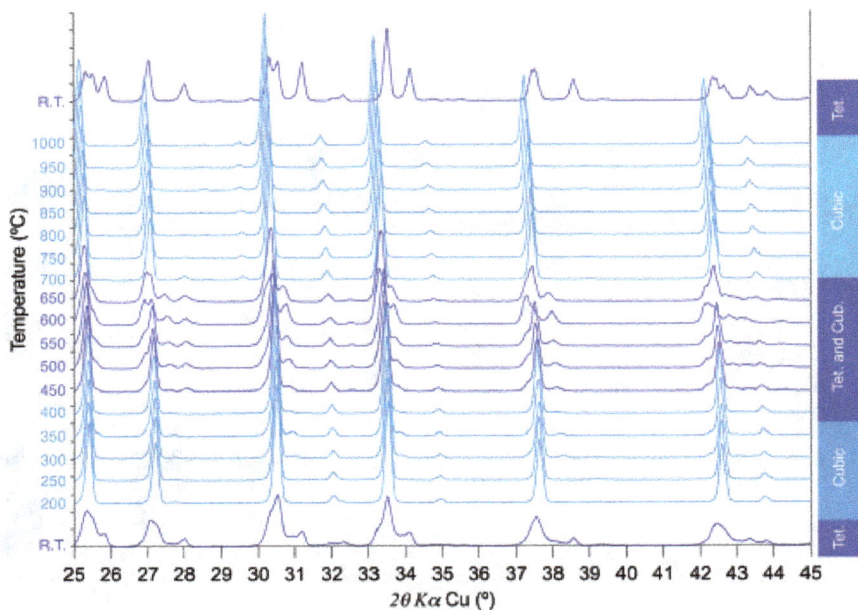

Fig. 5.24   HT-XRD patterns of tetragonal LLZ from 200°C to 1000°C.

Source: Reproduced from Ref. [117] with permission from Royal Society of Chemistry.

Finally, the tetragonal phase changed into the high-temperature cubic (HT-cubic) phase at 650°C. The Arrhenius plot of LLZ clearly showed a big gap in the bulk conductivity from 600°C to 650°C (Fig. 5.25). This corresponded to the phase transition temperature from tetragonal to HT-cubic phase. The tetragonal phase requires full Li-ion occupancy. Therefore, creation of Li vacancies is a good strategy to stabilize the high ionic conductive cubic phase at room temperature. By combining density-functional theory and molecular dynamics simulations, a critical Li vacancy concentration in the range of 0.4–0.5 per LLZ formula unit for stabilization of the cubic phase is necessary, irrespective of how the vacancies are introduced [119].To introduce the Li vacancies, doping with foreign multivalent ions like Al [7, 120–122], Ta [114, 123, 124], Ga [125, 126] Ge [127], Nb [128], Te [129], Y [13], Cr [131] and so on has been attempted and cubic

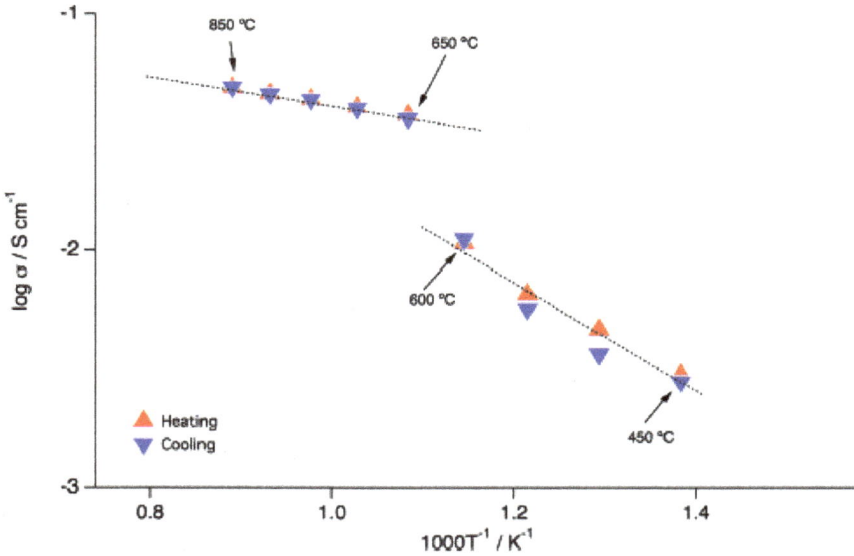

Fig. 5.25   Arrhenius plot of LLZ.

Source: Reproduced from Ref. [117] with permission from Royal Society of Chemistry.

LLZ was obtained successfully. In these dopants, Al, Ga and Ge are considered to be doped in Li sites and Ta, Y, Cr, Nb and Te sit in the Zr site, indicating that the doping of sites to create the Li vacancies does not relate to the stabilization of the cubic phase. The alkaline earth metals are generally thought to be confined exclusively to dodecahedral 8-co-ordinated sites (La sites) due to their large ionic radii. However, recent research has found that alkaline earth metals can also be doped at 6-co-ordinated sites (Zr sites), and thus superior ionic conductivity of $1.13 \times 10^{-3}\,\mathrm{S\,cm^{-1}}$ at room temperature is achieved [132, 133].

First-Principles studies also indicate that alkaline earth metals, i.e. Mg^{2+}, are stable at Zr site and can stabilize the cubic garnet phase [134]. The garnet framework has decent capability that ions with different valence state and different ionic radii can be doped into the lattice.

However, doping at the Li site may give rise to interference with the Li-ion conduction by the doped ions. Matsui *et al.* estimated the Li-ion conductivity of Al-free cubic phase at 25°C to be $1.71 \times 10^{-3} \, S \, cm^{-1}$ by extrapolating the Li-ion conductivity at 650–850°C [117]. This value is about one order higher than that of the Al-doped cubic phase because Al in the Li sites is thought to block Li-ion conduction. The maximum Li content in the cubic garnet is suggested to be 7.5 per formula by structural analysis using neutron diffraction and the highest conductivity would be obtained at Li content of 6.4 ± 0.1 per formula [135]. Also, the optimal lattice constant for the Li-ion conduction in $Li_{7+x-y}(La_{3-x}A_x)(Zr_{2-y}Nb_y)O_{12}$ (A = Alkaline metal) was suggested to be 12.94–12.96 Å (Fig. 5.26) [136].

Despite possessing several advantages over other types of conductors, reproducibility, especially with respect to Li-ion conductivity of LLZ solid electrolyte is still debated. For example, LLZ doped with Ga, Ta or Te exhibited high Li-ion conductivity of about $10^{-3} \, S \, cm^{-1}$ [9, 129, 137, 138]; However, lower conductivity value was also reported in similar compositions [125, 139, 140] This discrepancy would be explained by composition, sinterability and storage of sample. As mentioned above, Li content largely influences ionic

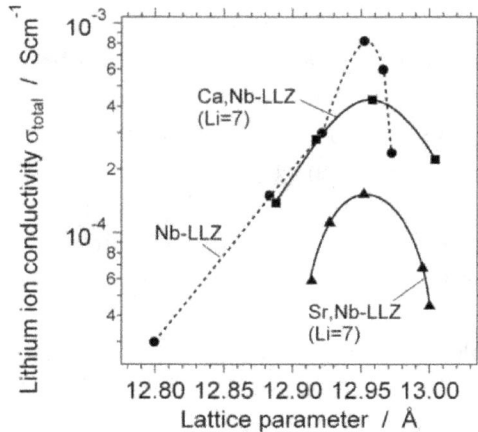

Fig. 5.26　Li-ion conductivities of Nb-LLZ, Sr, Nb-LLZ and Ca, Nb-LLZ at various lattice parameters [136].

conductivity of LLZ. In order to obtain the sintered body, normally LLZ powder is cold-pressed and sintered at high temperature above 1200°C. At this high temperature, Li evaporation from the sample is inevitable. Although it has been clarified that Li content in the sintered body is very sensitive to sintering temperature, Li sources and calcination temperature for LLZ powder preparation, precise control of Li content is very difficult [141–143]. Addition of excess Li in the starting precursor to compensate Li evaporation and/or covering the pelletized LLZ by mother powder to avoid Li evaporation usually have been adopted. However, the excess Li addition may promote Al^{3+} ion diffusion from the Al_2O_3 crucible in the sintering process, resulting in enhancement of sinterability and Li-ion conductivity of LLZ pellet [144]. The Al contamination cannot be avoided when Al_2O_3 crucible is used. However, it may be key for the stabilization of the high conductive cubic phase [120]. Additionally, Al_2O_3 is also considered to be a sintering additive which can increase relative density of the sintering body, leading to enhanced Li-ion conductivity [7]. Kotobuki *et al.* studied properties and electrochemical properties of Al_2O_3-added LLZ (Al-LLZ) in detail. In this study, 0.125 mol% of γ-Al_2O_3 was intentionally added. Cross-sections of the Al-LLZ pellet calcined at various temperatures demonstrated that the sinterability of the pellets increased with sintering temperature Grain boundaries could not be observed after sintering at 1100°C (Fig. 5.27). XRD reveals formation of impurity of $La_2Zr_2O_7$ Al-LLZ sintered at 1100°C, but the impurity of $La_2Zr_2O_7$ does not appear when it is sintered at 1000°C. Therefore, optimum sintering temperature should be around 1000°C. This sintering temperature is about 200°C lower than that of conventional LLZ sintering (>1200°C). The complex impedance plot of Al-LLZ is depicted in Fig. 5.28. A semicircle and Warburg impedance were observed at high and low frequency regions, respectively. Bulk and total Li-ion conductivities can be calculated intercepts at high and low frequency sides of the semicircle, respectively. The calculated Li-ion conductivities were 1.4×10^{-4} and $2.4 \times 10^{-4}\,S\,cm^{-1}$ for total and bulk contributions, respectively. The influence of stability of LLZ against Li metal by Al_2O_3 addition has been also studied. Figure 5.29

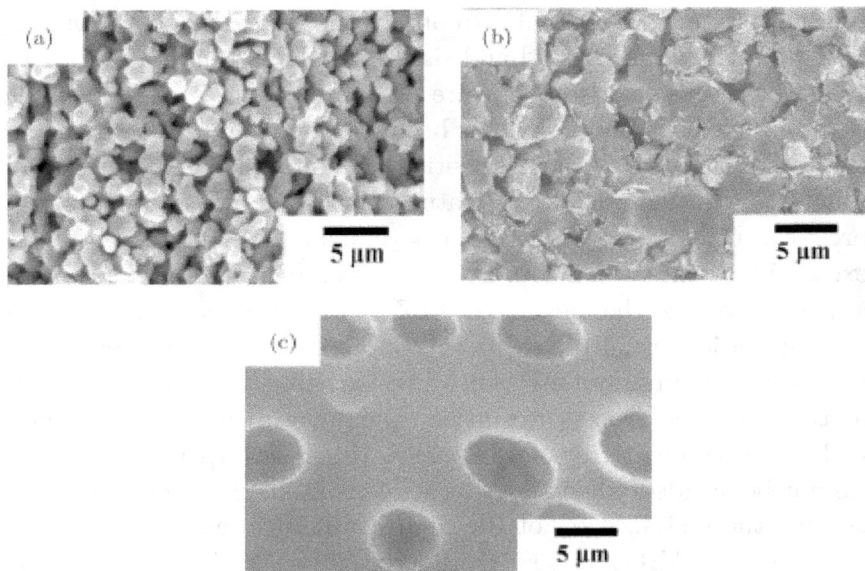

Fig. 5.27 SEM images of Al-LLZ calcined at (a) 900°C, (b) 1000°C and (c) 1100°C.

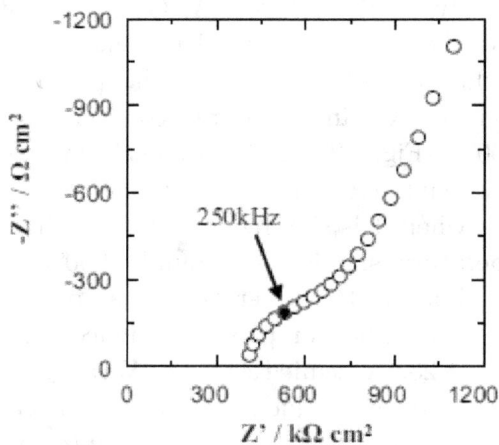

Fig. 5.28 Complex impedance plot of Al-LLZ.

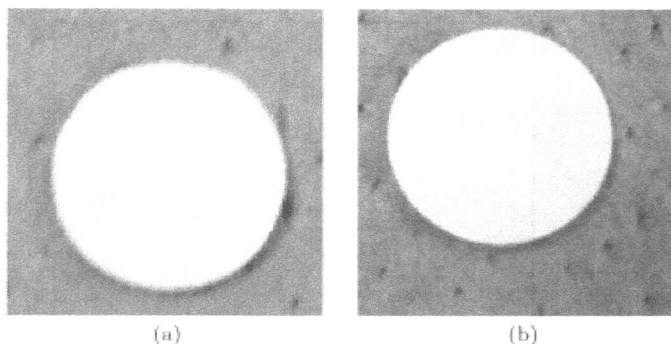

Fig. 5.29 Photos of Al-LLZ, (a) before and (b) after contact with molten Li for 72 h.

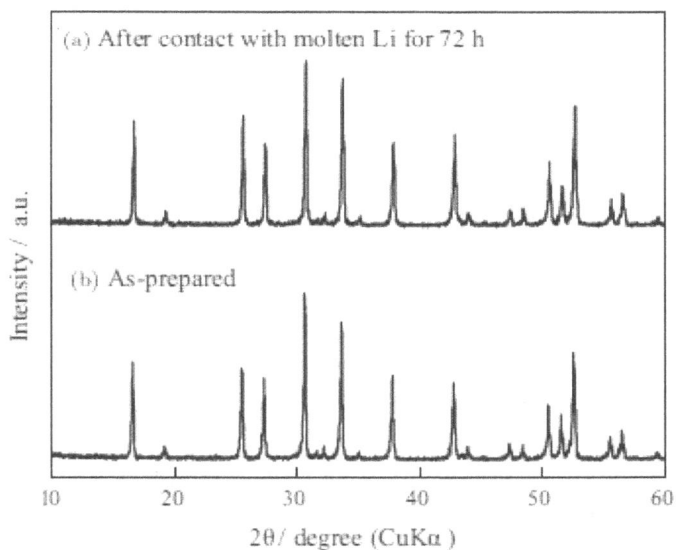

Fig. 5.30 XRD spectra of Al-LLZ before and after contact with molten Li for 72 h.

shows photographs of Al-LLZ before and after contact with molten Li for 72 h. No color change can be seen in Fig. 5.29, indicating that the Al$_2$O$_3$ addition did not influence the high stability of LLZ. This is also supported by XRD patterns of Al-LLZ before and after contact with molten Li (Fig. 5.30). There is no change in the patterns before

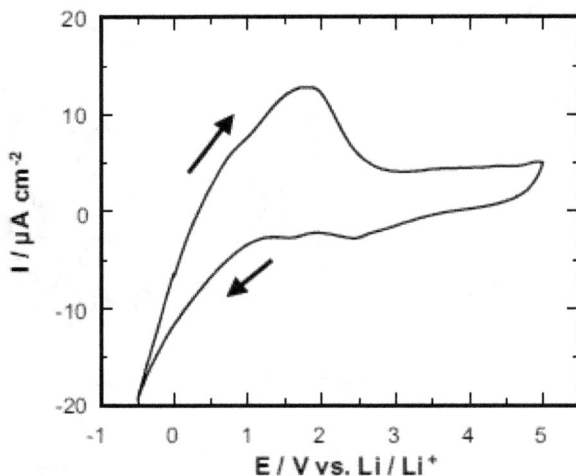

Fig. 5.31   Cyclic voltammogram of Li/Al–LLZ/Li cell.

and after contact with the molten Li. The Al-LLZ shows a wide electrochemical window at least above 5 V vs. Li^+/Li (Fig. 5.31).

It is hard to control exact amount of Al contamination into the LLZ lattice. It is also noted that Ga can play a similar role to increase the Li-ion conductivity [125] but must be added intentionally. However, the doping element may exist not only in the LLZ lattice but also at the grain boundary. Therefore, estimation of the doping amount in the LLZ lattice is very difficult, which is also true for other doping elements. The sintering process also influences the Li-ion conductivity of LLZ due mainly to enhancement of the grain-boundary conductivity. Baek *et al.* showed a high Li-ion conductivity of Ta-LLZ using Spark Plasma Sintering (SPS) [9]. Sakamoto's group used hot press technique for sintering LLZ and demonstrated a total conductivity of $2.3 \times 10^{-5}\,S\,cm^{-1}$ even in tetragonal LLZ [145], which is two orders higher than Awaka's value [10]. Recently, it was found that sintering atmosphere was also important to obtain high conductive pellet. $Li_{6.55}Ga_{0.15}La_3Zr_2O_{12}$ sintered in dry oxygen exhibited higher conductivity than that sintered in air [125, 126, 137, 146], which implies that LLZ is sensitive to ambient atmosphere [147] although LLZ was reported to be stable in air by Murugan *et al.* [6]. Li^+/H^+ exchange with adsorbed moisture occurs on the Li-stuffed garnet. As a result, Li-ion is released from the garnet lattice and forms LiOH, which further reacts with CO_2 and finally produces Li_2CO_3 [147,148]. Only the Li-stuffed garnets are sensitive to moisture [149–151], and the Li^+/H^+ exchange may proceed in the tetrahedral sites more than in the octahedral sites [152,153]. The sensitivity of LLZ to air, especially moisture and CO_2, also leads to complication of the tetragonal–cubic phase transition [116, 118, 154–156] and deviation of the composition, resulting in discrepancy of Li-ion conductivity of LLZ.

As mentioned above, many factors influence the Li-ion conductivity of LLZ, and these factors cause the reproducibility problem of LLZ. Establishment of preparation and storage procedures is key for further development of LLZ.

5.1.4 *Anti-perovskite type solid electrolyte*

Recently, a Li-rich anti-perovskite type solid electrolyte with a general formula of Li_3OX (X = Cl, Br) has been reported as a new family of solid electrolytes [157]. The anti-perovskite solid electrolyte was synthesized on the basis of $NaMgF_3$ and $(K, Na)MgF_3$ perovskite materials, which shows fast F^- conduction [158–161]. In the typical $A^+B^{2+}X_3^-$ perovskite, A is a monovalent cation like Na^+ and K^+, while B is a divalent cation like Mg^{2+}. X^- is a strongly electronegative monovalent anion like F^- and Cl^-. In the anti-perovskite structure, the electronegative anion, X^-, is replaced by the electropositive Li-ion and the monovalent and divalent cations are substituted by halogens and oxygen ions, respectively. The chemical formula of Li_3O (Cl, Br) can be described $A^-B^{2-}X_3^+$ in general. This formula is just an inversion of positions of cation and anions in the typical $A^+B^{2+}X_3^-$ perovskite. Therefore, this class of compounds is called "anti-perovskite solid electrolytes." Similar anti-perovskite solid electrolytes of lithium halide hydrates ($Li_{3-n}(OH_n)X$, n = 0.83–2, X = Br, Cl) were already reported in 2003 [162]. In these compounds, high conductive cubic phase ($\sim 10^{-4}\,S\,cm^{-1}$) appears above 33°C. However, the ionic conductivity of the non-cubic phase

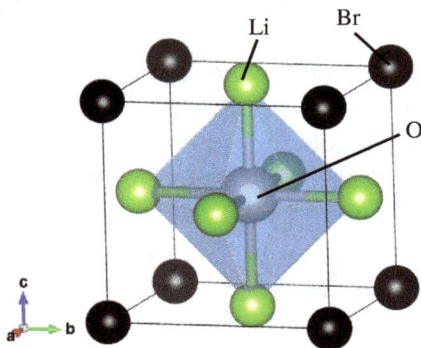

Fig. 5.32   Structure of anti-perovskite.

was below the detection limit ($< 10^{-8}\,\mathrm{S\,cm^{-1}}$), and this hinders the practical use of these materials. The anti-perovskite, Li_3OX contains 60 at.% of Li atoms in the perfect lattice. Therefore, Li_3OX is also called Li-rich anti-perovskite solid electrolyte.

The crystal structure of Li_3OBr anti-perovskite is depicted in Fig. 5.32 [163]. The Li-rich anti-perovskite belongs to the cubic lattice with space group of Pm-3m. The lattice constant was reported to be 4.02 Å. Li_3OCl also showed the same structure; however, lattice constant was a little smaller, 3.91 Å [157] due to smaller ion radius of Cl^-ion. The Li, halogen and O atoms are located at the octahedral vortices, cubic vortices and cubic centers, respectively [164]. In other words, the A site of perovskite is occupied by the halogen atoms, site B is occupied by oxygen atoms and the site for oxygen in the tradition perovskite is given to Li-ions. The oxygen ions are surrounded by 6 Li-ions forming Li_6O octahedra. In some perovskites, it is reported that the octahedral tilting along one particular lattice axial direction occurs in order to reduce the lattice energy due to structural disorder. The Li_3OCl and Li_3OBr are also thought to have the tilted octahedra.

The Li-ion conductivity of anti-perovskite is largely influenced by the sample preparation process. For example, Li-ion conductivity of as-prepared Li_3OCl, Li_3OBr and $Li_3OCl_{0.5}Br_{0.5}$ was low, $\sim 10^{-7}\,\mathrm{S\,cm^{-1}}$ after processing [157]. Upon annealing above 250°C for 24 h, Li-ion conductivity of these anti-perovskites shows a large improvement, by as much as 2 orders of magnitude [157]. With further increase in annealing temperature between 300°C and 360°C under vacuum for several days, high conductive of the anti-perovskite-structured pellets can be achieved. This improvement in conductivity may be caused by a formation of Li vacancies due to the annealing and/or due to Al^{3+} contamination as a result of a reaction with alumina crucible/substrate during high temperature treatment.

In fact, Schroeder *et al.* reported that a melt formed by heating a mixture of LiOH·OH and LiBr at 400°C was difficult to remove from alumina crucible in Li_3OBr preparation [165]. They quenched the melt onto a copper block to obtain Li_3OBr. Clear XRD diffraction peaks appeared in the quenched Li_3OBr sample, indicating that the quenched sample was crystal phase, not amorphous phase. Many compounds with the perovskite structure exhibit high ionic conductivity. Their perovskite structures which exhibit the best conductivity tend to be disordered, rather than ordered. For example, Ag_3SI which is also an anti-perovskite ion conductive ceramic, shows poor conductivity; however, the conductivity increases by about 2 orders of magnitude when S and I are disordered [166, 167].

Li-ions in the anti-perovskite move three dimensionally, as verified by theoretical calculations [168, 169]. However, the conduction mechanism is still debated widely. Zhang *et al.* have reported that the anti-perovskites with a perfect crystal structure would not be a good Li-ion conductor, and Li vacancies promote Li-ion migration by reducing the enthalpy barriers along the preferred pathways [168]. They also reported that Li sublattice melting is a reason for the observed superionic conductivity. This melting occurs well below the melting temperature of the anti-perovskite. The Li sublattice melting lowers the activation enthalpy for the Li-ion migration, and the superionic transport near the sublattice melting can explain the experimentally observed phenomena well. They concluded that Li vacancies and anion disorder caused by the sublattice melting were the primary mechanisms for superionic Li-ion conduction in the anti-perovskite materials. On the contrary, Emly *et al.* suggested interstitial dumbbell hop mechanism based on their first-principle

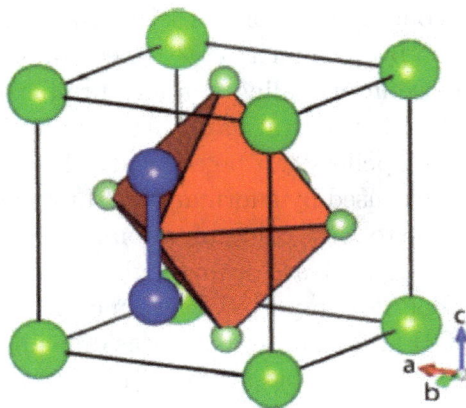

Fig. 5.33   Li dumbbell at the interstitial [168].

analysis (Fig. 5.33) [168]. There is an energetically stable Li dumbbell at the interstitium. Li-ions hop along the dumbbell in the perovskite crystal structure. The interstitial dumbbell hopping occurs at very low migration barrier in anti-perovskite Li_3OCl and Li_3OBr. Its migration barrier is roughly half as large as that for Li hopping into nearest neighbor vacancies reported (Fig. 5.34) [169]. This conduction mechanism involving the dumbbell interstitial is similar to the fluorine superionic perovskite, $KCaF_3$, predicted by MD (molecular dynamics) calculations [170]. However, the above two models cannot explain the superionic conductivity of anti-perovskites due to the high formation energy of Li interstitial defects. A work by Mouta *et al.* [171] also pointed out that the formation energy of Li Frenkel defect is much higher than that of other intrinsic charge neutral defects in Li_3OCl. Clarification of the formation mechanism of Li interstitial defects would be the key to gain more knowledge regarding the Li-ion conduction mechanism in the anti-perovskite.

Furthermore, the first-principle calculation has been used to determine the phase stability of anti-perovskites Li_3OCl and Li_3OBr. The anti-perovskites Li_3OCl and Li_3OBr are metastable compared with Li_2O, LiCl and LiBr at room temperature as well as at 0 K [168]. However, in considering the phase stability of anti-perovskites in the

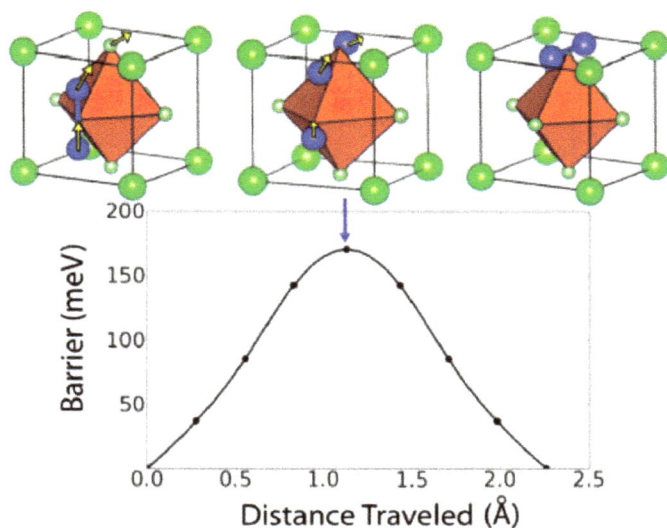

Fig. 5.34 Dumbbell hop mechanism [168].

absence of Li_2O, such as cases where Li_2O formation is kinetically difficult, the anti-perovskites Li_3OCl/Li_3OBr are stable and possess a variety of $LiCl$-/$LiBr$-deficient phases. The calculated electronic energy band gap is 6.44 eV for the base crystalline material of Li_3OCl [172]. Emly *et al.* also reported that the band gap of Li_3OCl exceeded 5 eV. However, they reported the metastable anti-perovskite becomes susceptible to decomposition into Li_2O_2, $LiCl$ and $LiClO_4$, over 2.5 V of an applied voltage, suggesting that Li_3OCl is suitable for low-voltage Li batteries [168].

In $Li_3OCl_{0.5}Br_{0.5}$ alloy, Cl–Br disorder results in high conductivity. A larger ion radius of Br^- is thought to narrow the channel of Li-ion diffusion. On the other hand, octahedral tilting occurs in the Li_3OCl. A partial substitution of Br to Cl prevents the octahedral tilting in the Li_3OCl by shrinking the channel size. A Cl-rich channel with Br-rich end points configuration leads to low vacancy migration barriers in the anti-perovskite structure. Incorporation of low levels of Br into Li_3OCl would increase the number of fast migration paths, but an excess of Br incorporation would cause

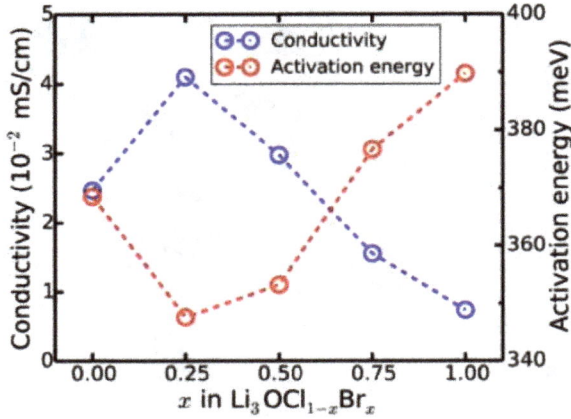

Fig. 5.35   Li ion conductivity and activation energy of $Li_3OCl_{1-x}Br_x$ alloy as a function of x.

"choking" in the channels, leading to decrease in conductivity. Deng *et al.* predicted the highest conductivity of $Li_3OCl_{1-x}Br_x$ structures would be obtained at $0.235 \leq x \leq 0.395$ (Fig. 5.35) [173]. The same behavior is observed in O^{2-} conducive perovskites, which have values equivalent to Li-ion conduction in the anti-perovskite. In Sr-doped $LaGaO_3$, the optimal O^{2-} conductivity is obtained with 20% Sr doping [174]. Interestingly, this value is similar to the prediction for highest conductivity of $Li_3OCl_{1-x}Br_x$. This increase of conductivity has been considered to be achieved by heteroatom substitution as well. Li vacancies can be created through doping with divalent atoms such as Mg and Ca. Introducing the divalent atoms at Li sites tends to form Li vacancies, which facilitate vacancy diffusion. Although the vacancies would not be as mobile as interstitial Li, the high concentration of vacancies may provide completely different migration mechanisms and contribute to the Li-ion conduction in the anti-perovskite, resulting in an increase of conductivity [168].

Not only Li-ion conductivity and conduction mechanisms, the stability of the Li-rich anti-perovskite type solid electrolyte in contact with Li metal was also studied. Lu *et al.* prepared Li_3OCl thin film by pulse laser deposition (PLD) [175] and tested the stability of the thin

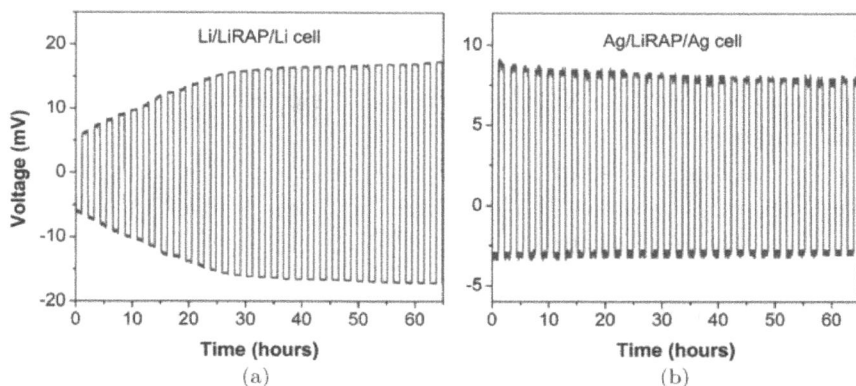

Fig. 5.36   Cyclability test of Li/Li$_3$ClO/Li cell.

Source: Reproduced from Ref. [175] with permission from Royal Society of Chemistry.

film in contact with Li metal. Figure 5.36 shows the cyclability test of a symmetric cell of Li/Li$_3$ClO/Li at a constant current of $100\,\mu$A at room temperature. The voltage increased with cycle, especially in the first 20 cycles and then became constant. Based on this observation, it was speculated that there were some interactions at the interfaces between Li metal and Li$_3$ClO. The interaction might be caused by oxidation of Li metal due to oxygen from Li$_3$ClO and/or other oxygen sources. The Li$_2$O formation as a result of oxidation of Li metal would increase cell resistance and voltage. After a certain number of cycles, the symmetric cell reached a stable condition. The authors called this behavior "self-stabilizing". This demonstrated that the Li metal could be used as the anode of all-solid-state batteries with Li$_3$ClO anti-perovskite solid electrolyte.

Another interesting study carried out by Schroeder *et al.* investigated the stability of Li$_3$OBr in common solvents using electrolytes such as DEC, DMC and PC with regard to application of the anti-perovskites for Li-air batteries.

In the Li-air batteries, a protective layer for the Li metal is needed to prevent its oxidation due to direct contact with air or liquid electrolytes. The Li$_3$OBr was found to be insoluble in these solvents, however, and lost its crystallinity in all solvents [165]. Additionally, the complete dissolution of Li_3OBr into 1.2M $LiPF_6$/70 wt.% EMC/30 wt.% EC electrolyte was observed after immersion for a few weeks.

Furthermore, a sodium-rich anti-perovskite, Na_3OBr, has also been recently reported. This sodium-rich anti-perovskite was prepared by mixing sodium oxide and NaBr, followed by calcination at 450°C for 24 h. The bulk and grain-boundary conductivities at 180°C were 9.02×10^{-7} and $1.29 \times 10^{-5}\,S\,cm^{-1}$, respectively [176]. The authors inferred that the formation energy of ion-diffusion-facilitating defects was much higher than that of Li analogues, resulting in a difference of ion conductivity among them.

The lithium-rich anti-perovskites are highly hygroscopic and should not be exposed to atmospheric moisture. This brings about a difficulty with regard to the handling of these materials, and hence this might be considered a shortcoming of the anti-perovskites. However, low toxicity LiOH and lithium halides are formed through decomposition in moisture. The anti-perovskites, therefore, are thought to be completely recyclable and eco-friendly. Additionally, the low cost of the starting materials is a preferred advantage when these materials will be mass produced [157].

5.2   Sulfide Li-Ion Conductive Ceramics

Because the high polarization of sulfide ions weakens the interaction between the anions and the lithium ions, sulfide solid electrolytes inherently tend to show fast ionic conduction. Additionally, the grain-boundary resistance of the sulfide electrolytes can be reduced even by cold-press [177].

Table 5.3 summarized the Li-ion conductivities of crystalline sulfide ceramics. When compared with the oxide counterparts, the sulfide ceramics possess about one order of magnitude higher conductivity. This has gained a lot of attention for development and application in solid electrolytes even though the sulfide ceramics react with ambient moisture and generate toxic H_2S gas [185]. Due to instability of the sulfide ceramics in air, especially in moisture,

Table 5.3 Li-ion conductivity of crystal sulfide solid electrolytes at room temperature

Electrolyte	Conductivity ($S\,cm^{-1}$)	Structure	Reference
$Li_{10}GeP_2S_{12}$	1.2×10^{-2}	—	[178]
$Li_{10}SnP_2S_{12}$	4×10^{-3}	—	[179]
Li_4GeS_4	2.0×10^{-7}	Thio-LISICON	[180]
$Li_{3.25}Ge_{0.25}P_{0.75}S_4$	2.2×10^{-3}	Thio-LISICON	[181]
Li_6PS_5Cl	1.3×10^{-3}	Argyrodite	[182]
Li_6PS_5I	2.2×10^{-4}	Argyrodite	[183]
Li_6PS_5Br	6.8×10^{-3}	Argyrodite	[184]

all production processes must be done in a dry condition. This is a major shortcoming of the sulfide ceramics with regard to their practical application.

5.2.1 *Thio-LISICON*

In 2000, Kanno *et al.* found a new sulfide type of Li-ion conductor, Li_4GeS_4, with γ-Li_3PO_4 structure [180]. This had a structure similar to that of LISICON (Li-ion super ionic conductor), such as $Li_{14}Zn(GeO_4)_4$ [186], which is one of the solid solutions of $Li_{2+2x}Zn_{1-x}GeO_4$ [187]. Therefore, it was named as thio-LISICON. Figure 5.37 depicts the crystal structure of Li_4GeS_4 [181]. Sulfur atoms are in a hexagonal close packing and the cations are distributed in various tetrahedral sites. Li can occupy Li1 (8d), Li2(4c) and Li3(8d) sites that are located between GeS_4 tetrahedra and form a 3D Li-ion conduction pathway. Although the pure solution phase of LISIOCN can be formed only in limited ranges of concentration and temperature [187], thio-LISICON can yield many solid solutions by cation substitution. The ionic conductivity of Li_4GeS_4 is quite low ($2.0 \times 10^{-7}\,S\,cm^{-1}$ at room temperature); however, the conductivity can be significantly increased by cation substitution.

For example, the $Li_{4+x+\delta}(Ge_{1-\delta'-x}Ga_x)S_4$ system exhibits the highest ionic conductivity of $6.5 \times 10^{-5}\,S\,cm^{-1}$ at $x = 0.25$. Additionally, the $Li_{4-x}Ge(Si,P)_{1-y}M_yS_4$ system (M = trivalent or pentavalent cation) presents an even higher ionic conductivity of $\sim 10^{-4}\,S\,cm^{-1}$. The highest conductivity of $2.2 \times 10^{-3}\,S\,cm^{-1}$ at 25°C

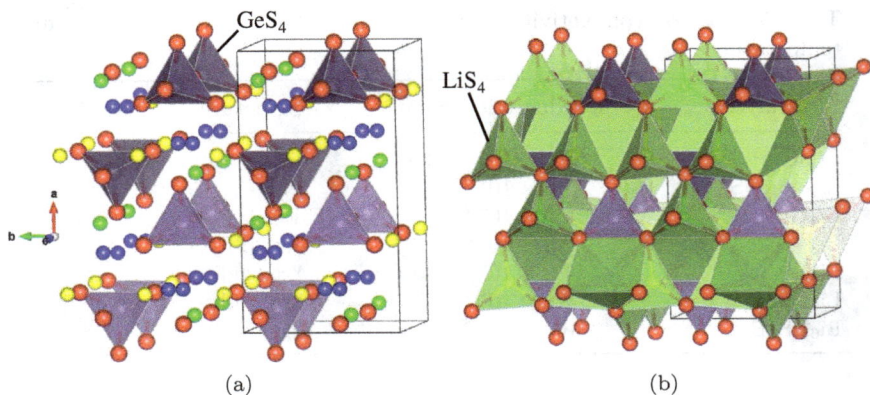

(a)   (b)

Fig. 5.37   Crystal structure of Li_4GeS_4. Both structures are same in which yellow: 8d (Li1) site, green: 4c (Li2) site, blue: 8d (Li3) site.

was obtained in $Li_{3.25}Ge_{0.25}P_{0.75}S_4$; however, this material is unstable in bulk form [181]. Li_4SiS_4-based thio-LISICON also reveals relatively high Li-ion conductivity ($6.4 \times 10^{-4}\,\mathrm{S\,cm^{-1}}$ at room temperature in $Li_{3.4}Si_{0.4}P_{0.6}S_4$) [188].

5.2.2 $Li_{10}GeP_2S_{12}$ *and related materials*

In 2011, Kamaya *et al.* reported that a novel crystalline sulfide Li-ion conductive ceramic, $Li_{10}GeP_2S_{12}$, exhibited conductivity of $1.2 \times 10^{-2}\,\mathrm{S\,cm^{-1}}$ at room temperature, which was the highest value so far among the inorganic solid electrolytes and comparable to that of the liquid electrolyte used in the commercial Li batteries [178]. The structure of this new superionic conductor was characterized by neutron diffraction. $Li_{10}GeP_2S_{12}$ has a tetragonal unit cell with space group of $P4_2/nmc$. The unit cell of $Li_{10}GeP_2S_{12}$ contains two tetrahedral sites: 4d and 2b sites. The 4d tetrahedral site is occupied by Ge and P ions, with an occupancy of about 0.5 for both atoms. On the contrary, the 2b tetrahedral site is occupied only by P ion. Li-ions sit in 16h, 4d and 8fsites (Fig. 5.38).

The 3D framework consists of $(Ge_{0.5}P_{0.5})S_4$ tetrahedra, PS_4 teterahedra, LiS_4 tetrahedra and LiS_6 octahedra (Fig. 5.39). The

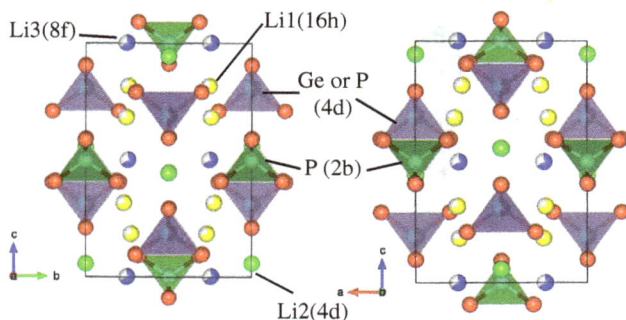

Fig. 5.38   Crystal structure of $Li_{10}GeP_2S_{12}$.

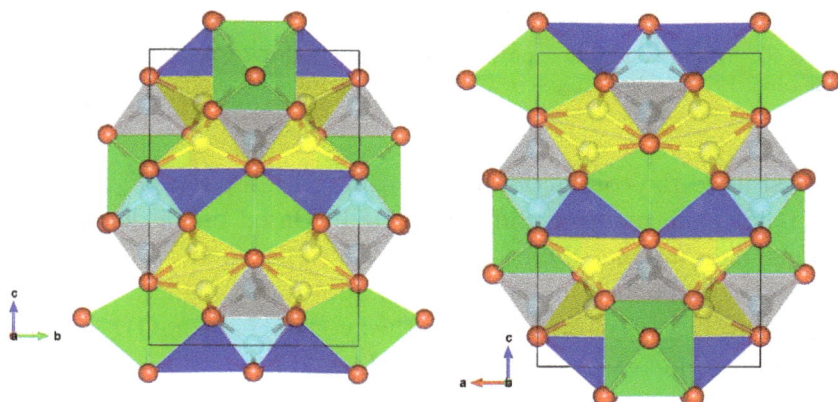

Fig. 5.39   Polyhedra of $Li_{10}GeP_2S_{12}$. Gray: $Ge_{0.5}P_{0.5}S_4$ tetrahedra, light blue: PS_4 tetrahedra, yellow: $Li(16h)S_4$ tetrahedra, green: $Li(4d)S_6$ octahedra, blue: $Li(8f)S_4$ tetrahedra.

Li-ion conduction in the $Li_{10}GeP_2S_{12}$ occurs 1D conduction along with c-axis.

Edge-shared $(Ge_{0.5}P_{0.5})S_4$ tetrahedra and LiS_6 octahedra align along with c-axis and form a 1D conduction pathway (Fig. 5.40). The 1D pathway is composed of LiS_4 tetrahedra in the 16h and 8f sites which share an edge. However, 3D conduction was also suggested as the basis of molecular dynamics simulations [179, 182]. Furthermore, single-crystal X-ray analysis showed that 3D motion of Li-ions plays an important role for the ionic conduction [189]. Kwon

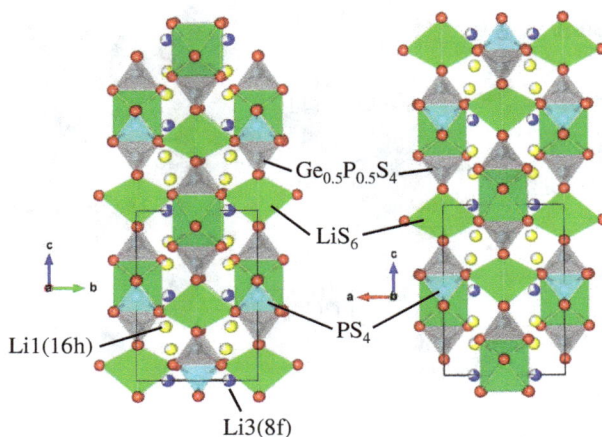

Fig. 5.40   Li ion conduction path of $Li_{10}GeP_2S_{12}$. Gray: $Ge_{0.5}P_{0.5}S_4$ tetrahedra, light blue: PS_4 tetrahedra, green: $Li(4d)S_6$ octahedra.

et al. studied the conduction mechanism of $Li_{10+\delta}Ge_{1+\delta}P_{2-\delta}S_{12}(0 \leq \delta \leq 0.35)$ using neutron diffraction [184]. Based on Li distribution, Li-ions migrate one dimensionally along the c-axis at 4.8–750 K. Also, 2D conduction in the ab plane appears at high temperature, suggesting that the conduction pathway changes from 1D to 3D as temperature increases. Furthermore, $Li_{10+\delta}Ge_{1+\delta}P_{2-\delta}S_{12}(\delta = 0.35)$ exhibits higher conductivity than $Li_{10}GeP_2S_{12}$, $1.42 \times 10^{-2}\,S\,cm^{-1}$ at 300 K. The activation energy of this material decreased with increase of temperature. This could be due to change the ion conduction pathway.

Kayama *et al.* also evaluated electrochemical stability of $Li_{10}GeP_2S_{12}$ using a $Li/Li_{10}GeP_2S_{12}/Au$ cell at a scan range of -0.5 to 5 V. No significant currents due to the electrolyte decomposition were observed in the scan range between -0.5 and 5 V. However, it was recently found that $Li_{10}GeP_2S_{12}$ was not stable when in contact with Li metal by *ab initio* calculations [179], cyclic voltammetry (CV) and *ex situ* X-ray diffraction (XRD) results [181]. Furthermore, elastic properties of $Li_{10}GeP_2S_{12}$, which are important for practical application for solid electrolytes, were systematically studied by *ab initio* calculations as well [190]. In fact, Kayama *et al.* fabricated

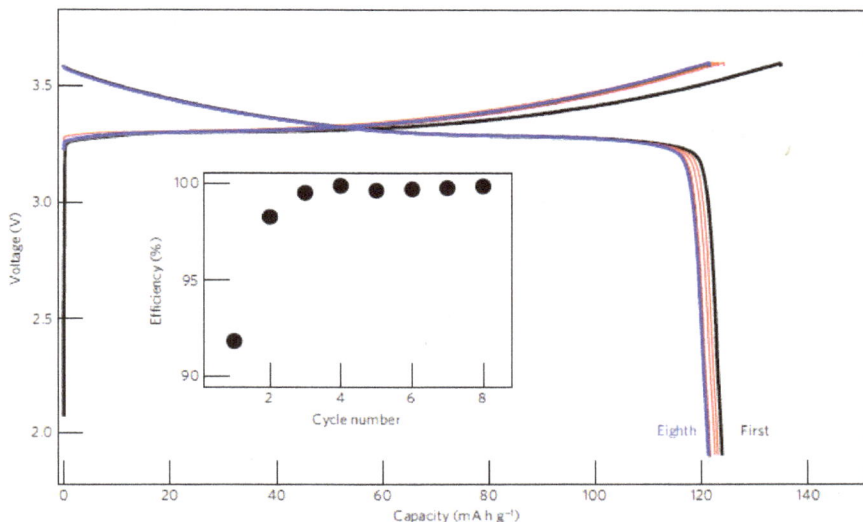

Fig. 5.41   Charge and discharge curves of $LiCoO_2/Li_{10}GeP_2S_{12}/In$ metal cell [178].

all-solid-state battery using $LiCoO_2$, $Li_{10}GeP_2S_{12}$ and Li metal as cathode, electrolyte and anode, respectively, and examined charge and discharge test [178]. Figure 5.41 shows the charge and discharge curves of $LiCoO_2/Li_{10}GeP_2S_{12}/In$ cell. Plateaus due to Li intercalation/deintercalation into/from $LiCoO_2$ were clearly observed in both charge and discharge curves. The discharge capacity is $120 \, mA \, h \, g^{-1}$ with Coulomb efficiency of almost 100% at the third cycle. Moreover, degradation of capacity and Coulomb efficiency was not observed until the eighth cycle. This proved that $Li_{10}GeP_2S_{12}$ solid electrolyte possessed high potential for application in all-solid-state battery.

Since the discovery of $Li_{10}GeP_2S_{12}$ in 2011, the $Li_{10}GeP_2S_{12}$ family has been intensively studied. Studies on the phase diagram, electrochemical stability and ionic conductivity of $Li_{10\pm1}MP_2X_{12}$ (M = Ge, Si, Sn, Al, P, X= O, S, Se) using first-principle calculations [191] reveal that isovalent substitution of Ge^{4+} with Si^{4+} and Sn^{4+} has a small effect on intrinsic properties. Aliovalent substitution does also not largely influence Li-ion conductivity. On the contrary, anion substitution has demonstrated significant effects on Li-ion conductivity. It was predicted that oxygen-substituted $Li_{10}MP_2O_{12}$ is not stable and has much lower Li-ion conductivity than that of their sulfide counterparts. Selenium-substituted compounds, such as $Li_{10}MP_2Se_{12}$, revealed slightly higher ionic conductivity, but lower electrochemical stability. The results with respect to the isovalent substitution (Si^{4+} and Sn^{4+}) have been authenticated by experiments [179, 192–194].

Recently, Kato *et al.* developed a novel superionic conductor in the $Li_{10}GeP_2S_{12}$ family, $Li_{9.54}Si_{1.74}P_{1.44}S_{11.7}Cl_{0.3}$, showing the highest Li-ion conductivity, $2.53 \times 10^{-2}\,S\,cm^{-1}$ at 25°C [195]. This value was twice that of the original $Li_{10}GeP_2S_{12}$ and the highest value reported among inorganic Li-ion conductors so far. The structure of $Li_{9.54}Si_{1.74}P_{1.44}S_{11.7}Cl_{0.3}$ is similar to $Li_{10}GeP_2S_{12}$, and it possesses 1D conduction pathway along with c-axis. Cl ion occupies 8g site, same as S ion in the $P(2b)X_4$ tetrahedra. The Cl ion in 8g site is thought to induce the 2D conduction pathway in the ab plane even at 25°C. This 3D conduction pathway leads to the extraordinarily high Li-ion conductivity of $Li_{9.54}Si_{1.74}P_{1.44}S_{11.7}Cl_{0.3}$. However, this material was confirmed to be electrochemically unstable toward Li metals like $Li_{10}GeP_2S_{12}$. Also, they reported in the same paper that the structure of $Li_{9.6}P_3S_{12}$ with $Li_{10}GeP_2S_{12}$ is more electrochemically stable. Therefore, it seems that the origin of electrochemical instability toward the Li metal of $Li_{9.54}Si_{1.74}P_{1.44}S_{11.7}Cl_{0.3}$ and $Li_{10}GeP_2S_{12}$ is not due to the crystal structure, but elements comprising the structure. The isovalent and/or aliovalent substitution of Ge^{4+} with more stable elements would lead to the development of novel superionic conductors of $Li_{10}GeP_2S_{12}$ family which are electrochemically stable toward the Li metal and have high Li-ion conductivity.

The discharge curves of two types of all-solid-state battery using $Li_{9.6}P_3S_{12}$ and $Li_{9.54}Si_{1.74}P_{1.44}S_{11.7}Cl_{0.3}$ as the electrolyte are shown in Fig. 5.42 [195]. The high voltage type of battery is configured by graphite/$Li_{9.6}P_3S_{12}$/$Li_{10}GeP_2S_{12}$/LiNbO$_3$-coated LiCoO$_2$ and the large-current type consists of $Li_4Ti_5O_{12}$/$Li_{9.54}Si_{1.74}P_{1.44}S_{11.7}Cl_{0.3}$/ LiNbO$_3$-coated LiCoO$_2$. The cells demonstrate excellent rate

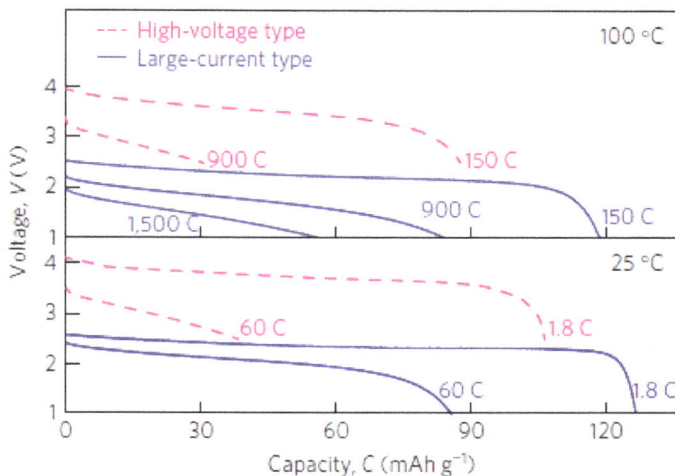

Fig. 5.42 Discharge curves of graphite/$Li_{9.6}P_3S_{12}$/$Li_{10}GeP_2S_{12}$/$LiNbO_3$-coated $LiCoO_2$ (high-voltage type) and $Li_4Ti_5O_{12}$/$Li_{9.54}Si_{1.74}P_{1.44}S_{11.7}Cl_{0.3}$/$LiNbO_3$-coated $LiCoO_2$ (large-current type) [195].

capability. Especially, the large-current type shows discharge behavior even at 60C at 25°C and at 1500C at 100°C. Additionally, the all-solid-state batteries also demonstrate excellent cycle performance under high current condition of 18 C at 100°C (Fig. 5.43). Conventional Li batteries with liquid electrolyte cannot be operated at this temperature due to evaporation of the electrolyte. These results clearly emphasize advantages of all-solid-state battery.

5.2.3 *Argyrodite*

In 2008, Li-ion conduction of lithium argyrodite, which is depicted with the general formula of Li_6PS_5X (X = Cl, Br, I), was found by Deiseroth *et al.* [196]. Argyrodites are a class of chalcogenide structures related to the mineral Ag_8GeS_6, which includes various fast Ag^+ or Cu^+ ion conductors such as $A_7PS_6(A = Ag^+, Cu^+)$ [197]. Because the ionic radii of Cu^+ and Li^+ are quite similar (Cu^+ = 74 pm, Li^+ = 73 pm at co-ordination number = 4), Cu ion would be substituted by Li-ion [198]. It is based on this concept that the lithium argyrodite has been synthesized. Li_6PS_5X (X = Cl, Br, I)is

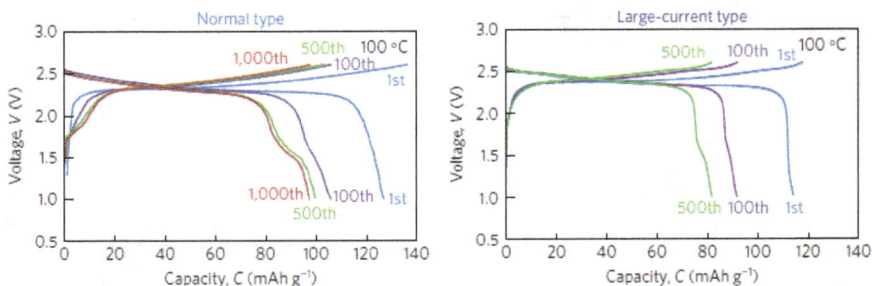

Fig. 5.43   Charge and discharge curves of $Li_4Ti_5O_{12}/Li_{10}GeP_2S_{12}/LiNbO_3$-coated $LiCoO_2$ (Normal type) and $Li_4Ti_5O_{12}/Li_{9.54}Si_{1.74}P_{1.44}S_{11.7}Cl_{0.3}/LiNbO_3$-coated $LiCoO_2$ (Large-current type) [195].

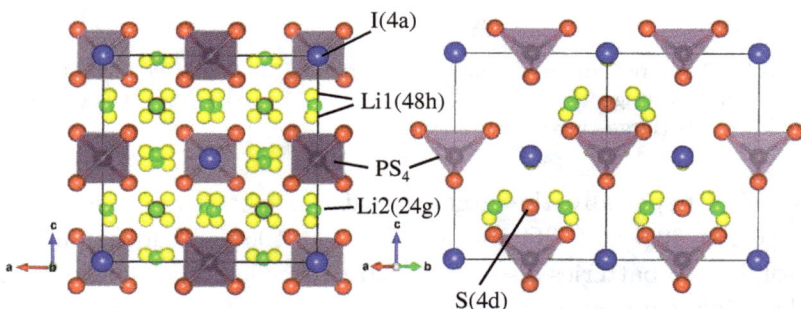

Fig. 5.44   Crystal structure of Li_6PS_5I.

a series of argyrodites, but the replacement of one chalcogen (S) with halogen atom (X) leads to different crystallographic positions from the original argyrodites, $A_7PS_6(A = Ag^+, Cu^+)$. Because of the replacement, the lithium argyrodite contains only six Li atoms per formula. The structures of three lithium argyrodites, Li_6PS_5I, Li_6PS_5Cl and Li_6PS_5Br, were characterized in detail by Rayavarapu *et al.* [183] using neutron diffraction. These three argyrodites have a slightly different in structure, especially with regard to anion ordering. Li_6PS_5I consists of a fully ordered arrangement of I ion on the 4a site (anion site 1), S ion on the 4d site (anion site 2) and tetrahedral PS_4 on the 4b site. Li-ions sit in the 48 h (Li(1)) and 24g (Li(2)) sites (Fig. 5.44).

Each Li(2) site is surrounded by two Li(1) sites. Since the distance between Li(1) and Li(2) sites is short (0.73696 Å), only one of these three sites is occupied, and this Li(1)–Li(2)–Li(1) series is considered as a single fully occupied site. On the contrary, anion disorder occurs in Li_6PS_5Cl and Li_6PS_5Br. The halide ions partially occupy the anion site 2. At same time, S ions also partially occupy the anion site 1. The halide ion occupancy in the anion site 2 is reported to be 0.63 and 0.43 for Li_6PS_5Cl and Li_6PS_5Br, respectively. Lighter halide ions seem to prefer the anion site 2. This anion disorder was also measured through NMR [196]. The measurement shows that Li_6PS_5I phase is fully ordered, Li_6PS_5Cl appears to be fully disordered, and Li_6PS_5Br is composed of both ordered and disordered crystallites or domains.

Li-ion conductivity of lithium argyrodite is expressed in the order of Br > Cl > I (Table 5.3). Anion disorder was observed in the lighter halides, which may induce high conductivity. The Li-ion in all the lithium argryrodites is thought to migrate three dimensionally. The Li-ion conduction pathway consists of a three dimensionally interconnected low-energy local pathway "cage" around anion sites 1 and 2. It is thought that the structure of these cages and their interconnectivity are different in different halide anions, resulting in diverse conductivity among the lithium agryrogates. However, the conduction pathway in the lithium argyrodites has not been well understood yet.

It is worthwhile to mention that the preparation procedure of the lithium agryrogates strongly influences the final products. Rayavarapu *et al.* prepared the lithium argyrodite using ball-mill mixing followed by annealing at 550°C for 5 h [183]. Their method was much shorter than original one proposed by Deiseroth *et al.* in which annealing was performed at 550°C for 7days. Additionally, Boulineau *et al.* reported a continuous ball-milling method that produced higher conductive samples than that produced by intermittent ball-milling method [182]. They also studied the effect of ball-milling duration on conductivity of the lithium argyrodite using Li_6PS_5Cl. A mixture of Li_2S, P_2S_5 and LiCl was ball-milled for different durations (1, 2, 5, 7.5, 10, 15 and 20 h). Only broad peaks of Li_2S and LiCl were observed on XRD pattern after milling for 1 and 2 h. After

5 h milling, Li_6PS_5Cl appeared with the remaining Li_2S. Almost pure Li_6PS_5Cl was obtained after 10 h milling. Conductivity was also influenced by the duration of ball-milling. The conductivity was enhanced from $2 \times 10^{-4}\,S\,cm^{-1}$ for 1 h milling to $1.33 \times 10^{-3}\,S\,cm^{-1}$ for 10 h milling, and then the conductivity decreased with further increase in ball-milling duration. The reason for the decrease of conductivity is still unclear. However, BET surface area of the ball-milled samples showed a similar trend to the conductivity. A coalescence of milled particle of Li_6PS_5Cl would decrease BET surface area and conductivity.

Another interesting behavior of the lithium argryrodites is the difference in conductivity between annealed and non-annealed samples. The Li-ion conductivity of ball-milled Li_6PS_5Cl, Li_6PS_5Br and Li_6PS_5I were found to be 3.3×10^{-5}, 3.2×10^{-5} and $2.2 \times 10^{-4}\,S\,cm^{-1}$, respectively. After annealing, the conductivity increased to 7.4×10^{-4} and $7.2 \times 10^{-4}\,S\,cm^{-1}$ for chloride and bromide samples, respectively. On the other hand, the conductivity of the iodide sample fell down to $4.6 \times 10^{-7}\,S\,cm^{-1}$, which was 500 times lower than that before annealing. The anion disorder is suggested a reason for this unusual behavior, but the conclusion is still widely debated.

References

[1] H. Aono, E. Sugimono, Y. Sadaoka, N. Imanaka, G. Adachi: Ionic conductivity of solid electrolytes based on lithium titanium phosphate, *J. Electrochem. Soc.* 137 (1990) 1023–1027.

[2] M. Kotobuki, M. Koishi: Sol–gel synthesis of $Li_{1.5}Al_{0.5}Ge_{1.5}(PO_4)_3$ solid electrolyte, *Ceram. Intl.* 41 (2015) 8562–8567.

[3] M. Ito, Y. Inaguma, W. H. Jung, L. Chen, T. Nakamura: High lithium ion conductivity in the perovskite-type compounds $Ln_{12}Li_{12}TiO_3$ (Ln = La,Pr,Nd,Sm), *Solid State Ionics* 70/71 (1994) 203–207.

[4] M. Kotobuki, Y. Suzuki, H. Munakata, K. Kanamura, Y. Sato, K. Yamamoto, T. Yoshida: Fabrication of three-dimensional battery using ceramic electrolyte with honeycomb structure by sol-gel process, *J. Electrochem. Soc.* 157(4) (2010) A493–A498.

[5] V. Thangadurai, H. Kaack, W. Weppner: Novel fast lithium ion conduction in garnet-type $Li_5La_3M_2O_{12}$ (M = Nb, Ta), *J. Am. Ceram. Soc.* 86(3) (2003) 437–440.

[6] R. Murugan, V. Thangadurai, W. Weppner: Fast lithium ion conduction in garnet-type $Li_7La_3Zr_2O_{12}$, *Angew. Chem. — Int. Ed.* 46 (2007) 7778–7781.

[7] M. Kotobuki, K. Kanamura, Y. Sato, T. Yoshida: Fabrication of all-solid-state lithium battery with lithium metal anode using Al_2O_3-added $Li_7La_3Zr_2O_{12}$ solid electrolyte, *J. Power Sources* 196 (2011) 7750–7754.

[8] M. Huang, W. Xu, Y. Shen, Y.-H. Lin, C.-W. Nan: X-ray absorption near-edge spectroscopy study on Ge-doped $Li_7La_3Zr_2O_{12}$: Enhanced ionic conductivity and defect chemistry, *Eelectrochimica Acta* 115 (2014) 581–586.

[9] S.-W. Baek, J.-M. Lee, Y. Kim, M.-S. Song, Y. Park: Garnet related lithium ion conductor processed by spark plasma sintering for all solid state batteries, *J. Power Sources* 249 (2014) 197–206.

[10] J. Awaka, N. Kijima, H. Hayakawa, J. Akimoto: Synthesis and structure analysis of tetragonal $Li_7La_3Zr_2O_{12}$ with the garnet-related type structure, *J. Solid State Chem.* 182 (2009) 2046–2052.

[11] M. Kotobuki, S. Song, R. Takahashi, S. Yanagiya, L. Lu: Improvement of Li ion conductivity of $Li_5La_3Ta_2O_{12}$ solid electrolyte by substitution of Ge for Ta, *J. Power Sources* 349 (2017) 105–110.

[12] J. B. Goodenough, H. Y. P. Hong, J. A. Kafalas: Fast Na^+-ion transport in skeleton structures, *Mater. Res. Bull.* 11(2) (1976) 203–220.

[13] M. Alami, R. Brochu, J. L. Soubeyroux, P. Gravereau, G. L. Flem, P. Hagenmuller: Structure and thermal expansion of $LiGe_2(PO_4)_3$, *J. Solid State Chem.* 90 (1991) 185–193.

[14] H. Aono, N. Imanaka, G.-Y. Adachi: High Li^+conducting ceramics, *Acc. Chem. Res.* 27(9) (1994) 265–270.

[15] V. Thangadurai, W. Weppner: Solid state lithium ion conductors: Design considerations by thermodynamic approach, *Ionics* 8 (2002) 281–292.

[16] N. Anantharamulu, K. K. Rao, G. Rambabu, B. V. Kumar, V. Radha, M. Vithal: A wide-ranging review on Nasicon type materials, *J. Mater. Sci.* 46 (2011) 2821–2837.

[17] F. Sudreau, D. Petit, J. P. Boilot: Dimorphism, phase transitions, and transport properties in $LiZr_2(PO_4)_3$, *J. Solid State Chem.* 83(1) (1989) 78–90.

[18] J. E. Iglesias, C. Pecharrom: Room temperature triclinic modification of NASICON-Type $LiZr_2(PO_4)_3$, *Solid State Ionics* 112(3–4) (1998) 309–318.

[19] M. Catti, S. Stramare, R. Ibberson: Lithium location in NASICON-type Li^+ conductors by neutron diffraction. I. Triclinic α-$LiZr_2(PO_4)_3$, *Solid State Ionics* 123 (1999) 173–180.

[20] K. Arbi, M. Ayadi-Trabelsi, J. Sanz: Li mobility in triclinic and rhombohedral phases of the NASICON-type compound $LiZr_2(PO_4)_3$ as deduced from NMR spectroscopy, *J. Mater. Chem.* 12 (2002) 2985–2990.

[21] M. Catti, A. Comotti, S. Di Blas: High-temperature lithium mobility in α-$LiZr_2(PO_4)_3$ NASICON by neutron diffraction, *Chem. Mater.* 15 (2003) 1628–1632.

[22] H. Xie, J. B. Goodenough, Y. Li: $Li_{1.2}Zr_{1.9}Ca_{0.1}(PO_4)_3$, a room-temperature Li-ion solid electrolyte, *J. Power Sources* 196 (2011) 7760–7762.

[23] Y. Li, M. Liu, K. Liu, C.-A. Wang: High Li^+ conduction in NASICON-type $Li_{1+x}Y_xZr_{2-x}(PO_4)_3$ at room temperature, *J. Power Sources* 240(0) (2013) 50–53.

[24] Y. Li, W. Zhou, X. Chen, X. Lu, Z. Cui, S. Xin, L. Xue, Q. Jia, J. B. Goodenough: Mastering the interface for advanced all-solid-state lithium rechargeable batteries. *Proc. Natl. Acad. Sci.* 7(2016) 13313–13317.

[25] K. Takada: Electrolytes: Solid oxide, *Enc. Electrochem. Power Sources* 5 (2009) 328–336.

[26] Z. X. Lin, H. J. Yu, S. C. Li, S. B. Tian: Phase relationship and electrical conductivity of $Li_{1+x}Ti_{2-x}Ga_xP_3O_{12}$ and $Li_{1+2x}Ti_{2-x}Mg_xP_3O_{12}$ systems, *Solid State Ionic* 18–19 (1986) 549–552.

[27] M. A. Subramanian, R. Subramaiun, A. Clearfield: Lithium ion conductors in the system $AB(IV)_2(PO_4)_3$ (B = Ti, Zr and Hf), *Solid State Ionic* 18–19 (1986) 562–569.

[28] Z. X. Lin, S. B. Tian: Phase relationship and electrical conductivity of $Na_3Zr_{2-x}Yb_xSi_{2-x}P_{1+x}O_{12}$ system, *Solid State Ionics* 9–10 (1983) 809–811.

[29] R. Sobiestianskas, A. Dindune, Z. Kanepe, J. Ronis, A. Kezionis, E. Kazakevicius, A Orliukas: Electrical properties of $Li_{1+x}Y_yTi_{2-y}(PO_4)_3$ ($x, y = 0.3; 0.4$) ceramics at high frequencies, *Mat. Sci. Eng. B* 76 (2000) 184–192.

[30] H. Aono, E. Sugimoto, Y. Sadaoka, N. Imanaka, G.-Y. Adachi: Ionic conductivity of the lithium titanium phosphate $(Li_{1+X}M_XTi_{2-X}(PO_4)_3,$ M = Al,Sc,Y,and La) systems, *J. Electrochem. Soc.* 136 (1989) 590–591.

[31] N. Bounar, A. Benabbas, F. Bouremmad, P. Ropa, J. C. Carru: Structure, microstructure and ionic conductivity of the solid solution $LiTi_{2-x}Sn_x(PO_4)_3$, *Physica B* 407 (2012) 403–407.

[32] I. A. Stenina, M. N. Kislitsyn, I. Y. Pinus, S. M. Haile, A. B. Yaroslavtsev: Phase transitions and ion conductivity in NASICON-type compounds $Li_{1\pm x}Zr_{2-x}M_x(PO_4)_3$, M = Ta, Nb, Y, Sc, In, *Defect Diffusion Forum* 249 (2006) 255–262.

[33] S.-C. Li, J. Cai, Z. X. Lin: Phase relationships and electrical conductivity of $Li_{1+x}Ge_{2-x}Al_xP_3O_{12}$ and $Li_{1+x}Ge_{2-x}Cr_xP_3O_{12}$ system, *Solid State Ionics* 28–30(Part 2) (1988) 1265–1270.

[34] H. Aono, E. Sugimoto, Y. Sadaoka, N. Imanaka, G. Y. Adachi: Electrical properties and sinterability for lithium germanium phosphate $Li_{1+x}M_xGe_{2-x}(PO_4)_3$, M = Al, Cr, Ga, Fe, Sc and In systems, *Bull. Chem. Soc. Jpn.* 65(8) (1992) 2200–2204.

[35] M. A. Paris, A. Martinez-Juarez, J. E. Iglesias, J. M. Rojo, J. Sanz: Phase transition and ionic mobility in $LiHf_2(PO_4)_3$ with NASICON structure, *Chem. Mater.* 9 (1997) 1430–1436.

[36] H. Aono, E. Sugimoto, Y. Sadaoka, N. Imanaka, G.-Y. Adachi: Electrical properties and crystal structure of solid electrolyte based on lithium hafnium phosphate $LiHf_2(PO_4)_3$, *Solid State Ionics* 62 (1993) 309–316.

[37] B. V. R. Chowdari, K. Radhakrishnan, K. A. Thomas, G. V. S. Rao: Ionic conductivity studies on $Li_{1-x}M_{2-x}M'_xP_3O_{12}$ (H = Hf, Zr; M' = Ti, Nb), *Mater. Res. Bull.* 24 (1989) 221–229.

[38] W. Zhao, L. Chen, R. Xue, J. Min, W. Cui: Ionic conductivity and luminescence of Eu^{3+}-doped $LiTi_2(PO_4)_3$, *Solid State Ionics* 70/71(Part 1) (1994) 144–146.

[39] S. Wang, L. Ben, H. Li, L. Chen: Identifying Li^+ion transport properties of aluminum doped lithium titanium phosphate solid electrolyte at wide temperature range, *Solid State Ionics* 268 (2014) 110–116.

[40] S. Hamdoune, D. Tran Qui, E. J. L. Schouler: Ionic conductivity and crystal structure of $Li_{1+x}Ti_{2-x}In_xP_3O_{12}$, *Solid State Ionics* 18–19 (1986) 587–591.

[41] S. Wong, P. J. Newman, A. S. Best, K. M. Nairn, D. R. MacFarlane, M. Forsyth: Towards elucidating microscopic structural changes in Li-ion conductors $Li_{1+y}Ti_{2-y}Al_y(PO_4)_3$ and $Li_{1+y}Ti_{2-y}Al_y(PO_4)_{3-x}$ $(MO_4)_x$(M = V and Nb): X-ray and Al-27 and P-31 NMR studies, *J. Mater. Chem.* 8(10) (1998) 2199–2203.

[42] C. J. Leo, G. V. S. Rao, B. V. R. Chowdari: Effect of MgO addition on the ionic conductivity of $LiGe_2(PO_4)_3$ ceramics, *Solid State Ionics* 159 (2003) 357–367.

[43] H. Aono, S. Eisuke, S. Yoshihiko, I. Nobuhito, G. Y. Adachi: Ionic conductivity of $LiTi_2(PO_4)_3$ mixed with lithium salts. *Chem. Lett.* 19 (1990) 331–334.

[44] H. Aono, S. Eisuke, S. Yoshihiko, I. Nobuhito, G. Y. Adachi: Electrical property and sinterability of $LiTi_2(PO_4)_3$ mixed with lithium salt (Li_3PO_4 or Li_3BO_3), *Solid State Ionics* 47(1991) 257–264.

[45] M. Zhang, K. Takahashi, N. Imanishi, Y. Takeda, O. Yamamoto, B. Chi, J. Pu, J. Li: Preparation and electrochemical properties of $Li_{1+x}Al_xGe_{2-x}(PO_4)_3$synthesized by a sol-gel method, *J. Electrochem. Soc.* 159(7) (2012) A1114–A1119.

[46] L. Huang, Z. Wen, M. Wu, X. Wu, Y. Liu, X. Wang: Electrochemical properties of $Li_{1.4}Al_{0.4}Ti_{1.6}(PO_4)_3$ synthesized by a co-precipitation method, *J. Power Sources* 196 (2011) 6943–6946.

[47] M. Kotobuki, M. Koishi, Y. Kato: Preparation of $Li_{1.5}Al_{0.5}Ti_{1.5}(PO_4)_3$ solid electrolyte via a co-precipitation method, *Ionics* 19 (2013) 1945–1948.

[48] S. Duluard, A. Paillassa, L. Puech, P. Vinatier, V. Turq, P. Rozier, P. Lenormand, P. L. Taberna, P. Simon, F. Ansart: Lithium conducting solid electrolyte $Li_{1.3}Al_{0.3}Ti_{1.7}(PO_4)_3$ obtained via solution chemistry, *J. Eur. Ceram. Soc.* 33 (2013) 1145–1153.

[49] X. Xu, Z. Wen, X. Yang, J. Zhang, Z. Gu: High lithium ion conductivity glass-ceramics in Li_2O-Al_2O_3-TiO_2-P_2O_5 from nanoscaled glassy powders by mechanical milling, *Solid State Ionics* 177 (2006) 2611–2615.

[50] C. Delmas, A. Nadiri, J. L. Soubeyroux: The Nasicon-type titanium phosphates $ATi_2(PO_4)_3$ (A = Li, Na) as electrode materials, *Solid State Ionics* 28–30(0) (1988) 419–423.

[51] A. Aatiq, M. Menetrier, L. Croguennec, E. Suard, C. Delmas: On the structure of $Li_3Ti_2(PO_4)_3$, *J. Mater. Chem.* 12(10) (2002) 2971–2978.

[52] C. J. Leo, B. V. R. Chowdari, G. V. Subha Rao, J. L. Souquet: Lithium conducting glass ceramic with Nasicon structure, *Mater. Res. Bull.* 37 (2002) 1419–1430.

[53] X. Xu, Z. Wen, X. Wu, X. Yang, Z. Gu: Lithium ion-conducting glass-ceramics of $Li_{1.5}Al_{0.5}Ge_{1.5}(PO_4)_3$-$xLi_2O$ (x = 0.0 − 0.20) with good electrical and electrochemical properties, *J. Am. Ceram. Soc.* 90 (2007) 2802–2806.

[54] J. K. Feng, L. Lu, M. O. Lai: Lithium storage capability of lithium ion conductor $Li_{1.5}Al_{0.5}Ge_{1.5}(PO_4)_3$, *J. Alloy. Compd.* 501(2) (2010) 255–258.

[55] H. Xie, Y. Li, J. B. Goodenough: NASICON-type $Li_{1+2x}Zr_{2-x}Ca_x(PO_4)_3$ with high ionic conductivity at room temperature, *RSC Adv.* 1(9) (2011) 1728–1731.

[56] J. Fu, Fast Li^+ ion conducting glass-ceramics in the system Li_2O–l_2O_3–GeO_2–P_2O_5, *Solid State Ionics* 104 (1997) 191–194.

[57] X. Xu, Z. Wen, J. Wu, X. Yang: Preparation and electrical properties of NASICON-type structured $Li_{1.4}Al_{0.4}Ti_{1.6}(PO_4)_3$ glass-ceramics by the citric acid-assisted sol–gel method, *Solid State Ionics* 178 (2007) 29–34.

[58] X. Xu, Z. Wen, X. Yang, L. Chen: Dense nanostructured solid electrolyte with high Li-ion conductivity by spark plasma sintering technique, *Mater. Res. Bull.* 43 (2008) 2334–2341.

[59] J. S. Thokchom, N. Gupta, B. Kumar: Superionic conductivity in a lithium aluminum germanium phosphate glass-ceramic, *J. Electrochem. Soc.* 155 (2008) A915–A920.

[60] H. Kun, Y. Wang, C. Zu, H. Zhao, Y. Liu, J. Chen, B. Han, J. Ma: Influence of Al_2O_3 additions on crystallization mechanism and conductivity of Li_2O–GeO_2–P_2O_5 glass-ceramics, *Physica B: Condensed Matter* 406 (2011) 3947–3950.

[61] A. Kubanska, L. Castro, L. Tortet, O. Schäf, Michaël Dollé, R. Bouchet: Elaboration of controlled size $Li_{0.5}Al_{0.5}Ge_{1.5}(PO_4)_3$ crystallites from glass-ceramics, *Solid State Ionics* 254 (2014) 44–50.

[62] J. Yang, Z. Huang, B. Huang, J. Zhou, X. Xu: Influence of phosphorus sources on lithium ion conducting performance in the system of Li_2O–GeO_2–P_2O_5 glass-ceramics, *Solid State Ionics* 269 (2015) 61–65.

[63] H. Yamamoto, M.Tabuchi, T. Takeuchi, H. Kageyama, O. Nakamura: Ionic conductivity enhancement in $LiGe_2(PO_4)_3$ solid electrolyte, *J. Power Sources* 68 (1997) 397–401.

[64] J. Brous, I. Fankuchen, E. Banks: Rare earth titanates with a perovskite structure, *Acta Crystallogr.* 6 (1953) 67–70.

[65] J. A. Alonso, J. Sanz, J. Santamaria, C. Leon, A. Varez, M. T. Fernandez-Diaz: On the location of Li^+ cations in the fast Li-cation conductor $La_{0.5}Li_{0.5}TiO_3$ perovskite, *Angew. Chem. Int. Ed.* 39(3) (2000) 619–621; H. Y.-P. Hong: Crystal structure and ionic conductivity of $Li_{14}Zn(GeO_4)_4$ and other new Li^+ superionic conductor, *Mater. Res. Bull.* 13 (1978) 117–124.

[66] Y. Inaguma, C. Liquan, M. Itoh, T. Nakamura, T. Uchida, H. Ikuta, M. Wakihara: High ionic-conductivity in lithium lanthanum titanate, *Solid State Commun.* 86(10) (1993) 689–693.

[67] A. G. Belous, G. N. Novitskaya, S. V. Polyanetskaya, Y. I. Gornikov: Study of complex oxides with the composition $La_{2/3-x}Li_{3x}TiO_3$, *Inorg. Mater.* 23 (1987) 412–415.

[68] S. Stramare, V. Thangadurai, W. Weppner: Lithium lanthanum titanates: A review, *Chem. Mater.* 15(21) (2003) 3974–3999.

[69] H. Kawai, J. Kuwano: Lithium ion conductivity of A-site deficient perovskite solid solution $La_{0.67x}Li_{3x}TiO_3$, *J. Electrochem. Soc.* 141(7) (1994) L78–L79.

[70] J. Ibarra, A. Varez, C. Leon, J. Santamaria, L. M. Torres-Martinez, J. Sanz: Influence of composition on the structure and conductivity of the fast ionic conductors $La_{2/3x}Li_{3x}TiO_3$ ($0.03 \leq x \leq 0.167$), *Solid State Ionics* 134(3–4) (2000) 219–228.

[71] B. S. Youmbi, S. Zekeng, S. Domngang, F. Calvayrac, A. Bulou: An *ab initio* molecular dynamics study of ionic conductivity in hexagonal lithium lanthanum titanate oxide $La_{0.5}Li_{0.5}TiO_3$, *Ionics* 18(4) (2012) 371–377.

[72] S. Garcia-Martin, U. Amador, A. Morata-Orrantia, J. Rodriguez-Carvajal, M. A. Alario-Franco: Structure, microstructure, composition and properties of lanthanum lithium titanates and some substituted analogues, *Zeitschrift fur Anorganische und Allgemeine Chemie* 635(15) (2009) 2363–2373.

[73] X. Gao, C. A. J. Fisher, T. Kimura, Y. H. Ikuhara, H. Moriwake, A. Kuwabara, H. Oki, T. Tojigamori, R. Huang, Y. Ikuhara: Lithium atom and a-site vacancy distributions in lanthanum lithium titanate, *Chem. Mater.* 25(9) (2013) 1607–1614.

[74] O. Bohnke, H. Duroy, J. L. Fourquet, S. Ronchetti, D. Mazza: In search of the cubic phase of the Li^+ ion-conducting perovskite $La_{2/3x}Li_{3x}TiO_3$: Structure and properties of quenched and *in situ* heated samples, *Solid State Ionics* 149(3–4) (2002) 217–226.

[75] A. Mei, X.-L. Wang, Y.-C. Feng, S.-J. Zhao, G.-J. Li, H.-X. Geng, Y.-H. Lin, C.-W. Nan: Enhanced ionic transport in lithium lanthanum titanium oxide solid state electrolyte by introducing silica, *Solid State Ionics* 179(39) (2008) 2255–2259.

[76] Y. Harada, Y. Hirakoso, H. Kawai, J. Kuwano: Order-disorder of the A-site ions and lithium ion conductivity in the perovskite solid solution $La_{0.67x}Li_{3x}TiO_3$ ($x = 0.11$), *Solid State Ionics* 121(1–4) (1999) 245–251.

[77] C. León, J. Santamaria, M. A. Paris, J. Sanz, J. Ibarra, L. M. Torres: Non-Arrhenius conductivity in the fast ionic conductor $Li_{0.5}La_{0.5}TiO_3$: Reconciling spin-lattice and electrical-conductivity relaxations, *Phys. Rev. B* 56 (1997) 5302.

[78] Y. Inaguma, L. Q. Chen, M. Itoh, T. Nakamura: Candidate compounds with perovskite structure for high lithium ionic conductivity, *Solid State Ionics* 70/71 (1994) 196–202.

[79] Y. Harada, T. Ishigaki, H. Kawai, J. Kuwano: Lithium ion conductivity of polycrystalline perovskite $La_{0.67x}Li_{3x}TiO_3$ with ordered and disordered arrangements of the A-site ions, *Solid State Ionics* 108(1–4) (1998) 407–413.

[80] A. Morata-Orrantia, S. García-Martín, Miguel Á. Alario-Franco: Optimization of lithium conductivity in La/Li titanates, *Chem. Mater.* 15 (2003) 3991–3995.

[81] T. Okumura, K. Yokoo, T. Fukutsuka, Y. Uchimoto, M. Saito, K. Amezawa: Improvement of Li-ion conductivity in A-site disordering lithium-lanthanum-titanate perovskite oxides by adding LiF in synthesis, *J. Power Sources* 189 (2009) 536–538.

[82] T. Teranishi, M. Yamamoto, H. Hayashi, A. Kishimoto: Lithium ion conductivity of Nd-doped (Li, La) TiO_3 ceramics, *Solid State Ionics* 243 (2013) 18–21.

[83] C. W. Ban, G. M. Choi: The effect of sintering on the grain boundary conductivity of lithium lanthanum titanates, *Solid State Ionics* 140(3–4) (2001) 285–292.

[84] H. Geng, A. Mei, Y. Lin, C. Nan: Effect of sintering atmosphere on ionic conduction and structure of $Li_{0.5}La_{0.5}TiO_3$ solid electrolytes, *Mater. Sci. Eng. B* 164(2) (2009) 91–95.

[85] C. Ma, K. Chen, C. Liang, C.-W. Nan, R. Ishikawa, K. More, M. Chi: Atomic-scale origin of the large grain-boundary resistance in perovskite Li-ion-conducting solid electrolytes, *Energy Environ. Sci.* 7(5) (2014) 1638–1642.

[86] F. Aguesse, J. M. Lopez del Amo, V. Roddatis, A. Aguadero, J. A. Kilner: Enhancement of the grain boundary conductivity in ceramic $Li_{0.34}La_{0.55}TiO_3$ electrolytes in a moisture-free processing environment, *Adv. Mater. Interfaces* 1 (2014) 1300143.

[87] Y. Inaguma, M. Nakashima: A rechargeable lithium-air battery using a lithium ion-conducting lanthanum lithium titanate ceramics as an electrolyte separator, *J. Power Sources* 228 (2013) 250–255.

[88] S. Kumazaki, Y. Iriyama, K.-H. Kim, R. Murugan, K. Tanabe, K. Yamamoto, T. Hirayama, Z. Ogumi: High lithium ion conductive $Li_7La_3Zr_2O_{12}$ by inclusion of both Al and Si, *Electrochem. Commun.* 13 (2011) 509–512.

[89] M. Kotobuki, H. Munakata, K. Kanamura, Y. Sato, T. Yoshida: Compatibility of $Li_7La_3Zr_2O_{12}$ solid electrolyte to all-solid-state battery using Li metal anode, *J. Electrochem. Soc.* 157(10) (2010) A1076–A1079.

[90] P. Birke, S. Scharner, R. A. Huggins, W. Weppner: Electrolytic stability limit and rapid lithium insertion in the fast-ion-conducting

$Li_{0.29}La_{0.57}TiO_3$ perovskite-type compound, *J. Electrochem. Soc.* 144(6) (1997) L167–L169.

[91] A. Kuhn, F. Garcia-Alvarado, A. Varez J. Sanz: Influence of percolation effects on lithium intercalation into $Li_{0.5-x}Na_xLa_{0.5}TiO_3$ ($0 \leq x \leq 0.5$) *Perovskites*, *J. Electrochem. Soc.* 152 (2005) A2285–A2290.

[92] H. M. Kasper: Series of rare earth garnets $Ln^{3+}{}_3M_2Li_3^+O_{12}$ (M = Te, W), *Inorg. Chem.* 8 (1969) 1000–1002.

[93] F. Abbattista, M. Vallino and D. Mazza: Remarks on binary system Li_2O–Me_2O_5 (Me = Nb, Ta), *Mater. Res. Bull.* 22 (1987) 1019–1027.

[94] H. Hyooma, K. Hayashi: Crystal structures of $La_3Li_5M_2O_{12}$ (M = Nb, Ta), *Mater. Res. Bull.* 23 (1988) 1399–1407.

[95] J. Isasi, M. L. Veiga, R. Saez-Puche, A. Jerez, C. Pico: Synthesis, structure determination and magnetic susceptibilities of the oxides Ln3Li5Sb2O12 (Ln $\neq$ Pr, Nd, Sm), *J. Alloys Compds.* 177 (1991) 251–257.

[96] J. Isasi, M. L. Veiga, A. Jerez, C. Pico: Synthesis and structure determination of a new phase in the La_2O_3–Li_2O–Sb_2O_5 system, *J. Less-Common Met.* 167 (1991) 381–385.

[97] V. Thangadurai, W. Weppner: Recent progress in solid oxide and lithium ion conducting electrolytes research, *Ionics* 12 (2006) 81–92.

[98] E. J. Cussen: The structure of lithium garnets: Cation disorder and clustering in a new family of fast Li^+ conductors, *Chem. Commun.* (2006) 412–413.

[99] R. Murugan, W. Weppner, P. Schmid-Beurmann, V. Thangadurai: Structure and lithium ion conductivity of bismuth containing lithium garnets $Li_5La_3Bi_2O_{12}$ and $Li_6SrLa_2Bi_2O_{12}$, *Mater. Sci. Eng. B* 143 (2007) 14–20.

[100] Y. X. Gao, X. P. Wang, W. G. Wang, Z. Zhuang, D. M. Zhang, Q. F. Fang: Synthesis, ionic conductivity, and chemical compatibility of garnet-like lithium ionic conductor $Li_5La_3Bi_2O_{12}$, *Solid State Ionics* 181 (2010) 1415–1419.

[101] R. Murugan, W. Weppner, P. Schmid-Beurmann, V. Thangadurai: Structure and lithium ion conductivity of garnet-like $Li_5La_3Sb_2O_{12}$ and $Li_6SrLa_2Sb_2O_{12}$, *Mater. Res. Bull.* 43(2008) 2579–2591.

[102] E. J. Cussen, T. W. S. Yip: A neutron diffraction study of the d^0 and d^{10} lithium garnets $Li_3Nd_3W_2O_{12}$ and $Li_5La_3Sb_2O_{12}$, *J. Solid State Chem.* 180(2007) 1832–1839.

[103] V. Thangadurai, W. Weppner: Effect of sintering on the ionic conductivity of garnet-related structure $Li_5La_3Nb_2O_{12}$ and In- and K-doped $Li_5La_3Nb_2O_{12}$, *J. Solid State Chem.* 179 (2006) 974–984.

[104] I. P. Roof, M. D. Smith, E. J. Cussen, H. C. zur Loye: Crystal growth of a series of lithium garnets $Ln_3Li_5Ta_2O_{12}$ (Ln = La, Pr, Nd): Structural properties, Alexandrite effect and unusual ionic conductivity, *J. Solid State Chem.* 182(2009) 295–300.

[105] R. Murugan, V. Thangadurai, W. Weppner: Effect of lithium ion content on the lithium ion conductivity of the garnet-like structure $Li_{5+x}BaLa_2Ta_2O_{11.5+0.5x}$ ($x = 0 - 2$), *Appl. Phys. A: Mater. Sci. Process.* 91(2008) 615–620.

[106] V. Thangadurai, W. Weppner: $Li_6ALa_2Nb_2O_{12}$ (A = Ca, Sr, Ba): A New class of fast lithium ion conductors with garnet-like structure, *J. Am. Ceram. Soc.* 88 (2005) 411–418.

[107] V. Thangadurai, W. Weppner: $Li_6ALa_2Ta_2O_{12}$ (A = Sr, Ba): Novel garnet-like oxides for fast lithium ion conduction, *Adv. Funct. Mater.* 15(1) (2005) 107–112.

[108] V. Thangadurai, S. Narayanan, D. Pinzaru: Garnet-type solid-state fast Li ion conductors for Li batteries: Critical review, *Chem. Soc. Rev.* 43 (2014) 4714–4727.

[109] M. Nakayama, M. Kotobuki, H. Munakata, M. Nogami, K. Kanamura: First-principles density functional calculation of electrochemical stability of fast Li ion conducting garnet-type oxides, *Phys. Chem. Chem. Phys.* 14 (2012) 10008–10014.

[110] M. Kotobuki, K. Kanamura: Fabrication of all-solid-state battery using $Li_5La_3Ta_2O_{12}$ceramic electrolyte, *Ceram. Intl.* 39 (2013) 6481–6487.

[111] M. Kotobuki, M. Koishi: Preparation of $Li_7La_3Zr_2O_{12}$ solid electrolyte via a sol-gel method, *Ceram. Intl.* 40 (2014) 5043–5047.

[112] R. Murugan, W. Weppner, P. Schmid-Beurmann, V. Thangadurai: Structure and lithium ion conductivity of bismuth containing lithium garnets $Li_5La_3Bi_2O_{12}$ and $Li_6SrLa_2Bi_2O_{12}$, *Mat. Sci. Eng. B — Solid* 143(1–3) (2007) 14–20.

[113] Y. Ren, K. Chen, R. Chen, T, Liu, Y. Zhang, C.-W. Nan: Oxide electrolytes for lithium batteries, *J. Am. Ceram. Soc.* 98 (2015) 3603–3623.

[114] A. Logeat, T. Kohler, U. Eisele, B. Stiaszny, A. Harzer, M. Tovar, A. Senyshyn, H. Ehrenberg, B. Kozinsky: From order to disorder: the structure of lithium conducting garnets $Li_{7-x}La_3Ta_xZr_{2-x}O_{12}$ ($x = 0-2$), *Solid State Ionics* 206(0) (2012) 33–38.

[115] J. Awaka, A. Takashima, K. Kataoka, N. Kijima, Y. Idemoto, J. Akimoto: Crystal structure of fast lithium-ion-conducting cubic $Li_7La_3Zr_2O_{12}$, *Chem. Lett.* 40 (2011) 60–62.

[116] G. Larraza, A. Orera, M. L. Sanjuan: Cubic phases of garnet-type $Li_7La_3Zr_2O_{12}$: The role of hydration, *J. Mater. Chem. A* 1 (2013) 11419–11428.

[117] M. Matsui, K. Takahashi, K. Sakamoto, A. Hirano, Y. Takeda, O. Yamamoto, N. Imanishi: Phase stability of a granet-type lithium ion conductor $Li_7La_3Zr_2O_{12}$, *Dalton Trans.* 43 (2014) 1019–1024.

[118] M. Matsui, K. Sakamoto, K. Takahashi, A. Hirano, Y. Takeda, O. Yamamoto, N. Imanishi: Phase transformation of the garnet structured lithium ion conductor: $Li_7La_3Zr_2O_{12}$, *Solid State Ionics* 262 (2014) 155–159.

[119] N. Bernstein, M. D. Johannes, K. Hoang: Origin of the structural phase transition in $Li_7La_3Zr_2O_{12}$, *Phys. Rev. Lett.* 109(20) (2012) 205702 5pp.

[120] C. A. Geiger, E. Alekseev, B. Lazic, M. Fisch, T. Armbruster, R. Langner, M. Fechtelkord, N. Kim, T. Pettke, W. J. F. Weppner: Crystal chemistry and stability of "$Li_7La_3Zr_2O_{12}$" garnet: A fast lithium-ion conductor, *Inorg. Chem.* 50 (2011) 1089–1097.

[121] A. Duvel, A. Kuhn, L. Robben, M. Wilkening, P. Heitjans: Mechanosynthesis of solid electrolytes: Preparation, characterization, and li ion transport properties of garnet-type Al-doped $Li_7La_3Zr_2O_{12}$crystallizing with cubic symmetry, *J. Phys. Chem. C* 116 (2012) 15192–15202.

[122] R. Takano, K. Tadanaga, A. Hayashi, M. Tatsumisago: Low temperature synthesis of Al-doped $Li_7La_3Zr_2O_{12}$ solid electrolyte by a sol–gel process, *Solid State Ionics* 255 (2014) 104–107.

[123] J. L. Allen, J. Wolfenstine, E. Rangasamy, J. Sakamoto: Effect of substitution (Ta, Al, Ga) on the conductivity of $Li_7La_3Zr_2O_{12}$, *J. Power Sources* 206 (2012) 315–319.

[124] Y. Li, C.-A. Wang, H. Xie, J. Cheng, J. B. Goodenough: High lithium ion conduction in garnet-type $Li_6La_3ZrTaO_{12}$, *Electrochem. Commun.* 13 (2011) 1289–1292.

[125] H. El Shinawi, J. Janek: Stabilization of cubic lithium-stuffed garnets of type "$Li_7La_3Zr_2O_{12}$" by addition of gallium, *J. Power Sources* 225 (2013) 13–19.

[126] M. A. Howard, O. Clemens, E. Kendrick, K. S. Knight, D. C. Apperley, P. A. Anderson, P. R. Slater: Effect of Ga incorporation on the structure and Li ion conductivity of $La_3Zr_2Li_7O_{12}$, *Dalton Trans.* 41 (2012) 12048–12053.

[127] M. Huang, A. Dumon, C.-W. Nan: Effect of Si, In and Ge Doping on High Ionic Conductivity of $Li_7La_3Zr_2O_{12}$, *Electrochem. Commun.* 21(0) (2012) 62–64.

[128] S. Ohta, T. Kobayashi, T. Asaoka: High Lithium Ionic Conductivity in the Garnet-Type Oxide $Li_{7-x}La_3(Zr_{2-x}, Nb_x)O_{12}$ ($x = 0 - 2$), *J. Power Sources* 196(6) (2011) 3342–3345.

[129] C. Deviannapoorani, L. Dhivya, S. Ramakumar, R. Murugan: Lithium ion transport properties of high conductive tellurium substituted $Li_7La_3Zr_2O_{12}$ Cubic lithium garnets, *J. Power Sources* 240(0) (2013) 18–25.

[130] R. Murugan, S. Ramakumar, N. Janani: High conductive yttrium doped $Li_7La_3Zr_2O_{12}$cubic lithium garnet, *Electrochem. Commun.* 13(12) (2011) 1373–1375.

[131] S. Song, B. Yan, F. Zheng, H. M. Duong, L. Lu: Crystal Structure, Migration mechanism and electrochemical performance of Cr-stabilized garnet, *Solid State Ionics* 268 (2014) 135–139; P. G. Bruce, A. R. West: Phase diagram of the LISICON, solid electrolyte system, Li_4GeO_4–Zn_2GeO_4, *Mater. Res. Bull.* 15 (1980) 379–385.

[132] S. F. Song, D. Sheptyakov, A. M. Korsunsky, H. M. Duong, L. Lu: High Li ion conductivity in a garnet-type solid electrolyte via unusual site occupation of the doping Ca ions, *Mater. Des.* 93 (2016) 232–237.

[133] S. F. Song, M. Kotobuki, F. Zheng, C. Xu, Y. Wang, W. Li, N. Hu, L. Lu: Roles of alkaline earth ions in garnet-type superionic conductors, *Chem Electro Chem* 4 (201) 266–271.

[134] J. Miara, W. D. Richards, Y. E. Wang, G. Ceder: First-principles studies on cation dopants and electrolyte cathode interphases for lithium garnets, *Chem. Mater.* 27 (2015) 4040–4047.

[135] H. Xie, J. A. Alonso, Y. Li, M. T. Fernandez-Diaz, J. B. Goodenough: Lithium Distribution in Aluminum-Free Cubic $Li_7La_3Zr_2O_{12}$, *Chem. Mater.* 23(16) (2011) 3587–3589.

[136] Y. Kihira, S. Ohta, H. Imagawa, T. Asaoka: Effect of simultaneous substitution of alkali earth metals and Nb in $Li_7La_3Zr_2O_{12}$ on lithium-ion conductivity, *ECS Electrochem. Lett.* 2(7) (2013) A56–A59.

[137] C. Bernuy-Lopez, W. Manalastas Jr., J. M. Lopez del Amo, A. Aguadero, F. Aguesse, J. A. Kilner: Atmosphere Controlled processing of Ga-substituted garnets for high li-ion conductivity ceramics, *Chem. Mater.* 26(12) (2014) 3610–3617.

[138] Y. Li, J.-T. Han, C.-A. Wang, H. Xie, J. B. Goodenough: Optimizing Li^+ conductivity in a garnet framework, *J. Mater. Chem.* 22(30) (2012) 15357–15361.

[139] Y. Wang, W. Lai: High ionic conductivity lithium garnet oxides of $Li_{7-x}La_3Zr_{2-x}Ta_xO_{12}$ compositions, *Electrochem. Solid-State Lett.* 15(5) (2012) A68–A71.

[140] D. Wang, G. Zhong, O. Dolotko, Y. Li, M. J. McDonald, J. Mi, R. Fu, Y. Yang: The synergistic effects of Al and Te on the structure and Li^+-mobility of garnet-type solid electrolytes, *J. Mater. Chem. A* 2(47) (2014) 20271–20279.

[141] Y. Ren, H. Deng, R. Chen, Y. Shen, Y. Lin, C.-W. Nan: Effects of Li source on microstructure and ionic conductivity of Al-contained $Li_{6.75}La_3Zr_{1.75}Ta_{0.25}O_{12}$ceramics, *J. Eur. Ceram. Soc.* 35(2) (2015) 561–572.

[142] M. Huang, T. Liu, Y. Deng, H. Geng, Y. Shen, Y. Lin, C.-W. Nan: Effect of sintering temperature on structure and ionic conductivity of $Li_{7-x}La_3Zr_2O_{12-0.5x}(x = 0.5-0.7)$ ceramics, *Solid State Ionics* 204–205(0) (2011) 41–45.

[143] R.-J. Chen, M. Huang, W.-Z. Huang, Y. Shen, Y.-H. Lin, C.-W. Nan: Effect of calcining and Al doping on structure and conductivity of $Li_7La_3Zr_2O_{12}$, *Solid State Ionics* 265(0) (2014) 7–12.

[144] K. Liu, J.-T. Ma, C.-A. Wang: Excess lithium salt functions more than compensating for lithium loss when synthesizing $Li_{6.5}La_3Ta_{0.5}Zr_{1.5}O_{12}$ in alumina crucible, *J. Power Sources* 260(0) (2014) 109–114.

[145] J. Wolfenstine, E. Rangasamy, J. L. Allen, J. Sakamoto: High conductivity of dense tetragonal $Li_7La_3Zr_2O_{12}$, *J. Power Sources* 208 (2012) 193–196.

[146] J. Wolfenstine, J. Ratchford, E. Rangasamy, J. Sakamoto, J. L. Allen: Synthesis and high Li-ion conductivity of Ga-stabilized cubic$Li_7La_3Zr_2O_{12}$, *Mater. Chem. Phys.* 134(2–3) (2012) 571–575.

[147] X.-P. W. W.-G. Wang, Y.-X. Gao, J.-F. Yang, Q.-F. Fang: Investigation on the stability of $Li_5La_3Ta_2O_{12}$lithium ionic conductors in humid environment, *Front. Mater. Sci.* 4(2) (2010) 189–192.

[148] C. Galven, J.-L. Fourquet, M.-P. Crosnier-Lopez, F. Le Berre: Instability of the lithium garnet $Li_7La_3Sn_2O_{12}$: Li^+/H^+ exchange and structural study, *Chem. Mater.* 23(7)(2011) 1892–1900.

[149] C. Galven, J. Dittmer, E. Suard, F. Le Berre, M.-P. Crosnier-Lopez: Instability of lithium garnets against moisture. Structural characterization and dynamics of $Li_{7-x}H_xLa_3Sn_2O_{12}$ and $Li_{5-x}H_xLa_3Nb_2O_{12}$, *Chem. Mater.* 24[17] (2012) 3335–3345.

[150] T. Luna, V. Thangadurai: Soft-chemistry of garnet-type $Li_{5+x}Ba_x La_{3-x}Nb_2O_{12}$ (x = 0, 0.5, 1): Reversible $H^+ \longleftrightarrow Li^+$ ion-exchange reaction and their X-ray, Li-7 MAS NMR, IR, and AC impedance spectroscopy characterization, *Chem. Mater.* 23(17) (2011) 3970–3977.

[151] T. Lina, V. Thangadurai: First Total H^+/Li^+ Ion exchange in garnet-type $Li_5La_3Nb_2O_{12}$ using organic acids and studies on the effect of Listuffing, *Inorg. Chem.* 51(3) (2012) 1222–1224.

[152] C. Li-quan, W. Lian-zhong, C. Guang-can, W. Gang, L. Z-rong: Investigation of new lithium ionic conductors $Li_{3+x}V_{1-x}Si_xO_4$, *Solid State Ionics* 9–10(1983) 149–152.

[153] A. Khorassani, A. R. West: Li^+ Ion conductivity in the system Li_4SiO_4–Li_3VO_4, *J. Solid State Chem.* 53(3) (1984) 369–375.

[154] S. Toda, K. Ishiguro, Y. Shimonishi, A. Hirano, Y. Takeda, O. Yamamoto, N. Imanishi: Low temperature cubic garnet-type CO_2-doped $Li_7La_3Zr_2O_{12}$, *Solid State Ionics* 233(0) (2013) 102–106.

[155] X. P. Wang, Y. Xia, J. Hu, Y. P. Xia, Z. Zhuang, L. J. Guo, H. Lu, T. Zhang, Q. F. Fang: Phase transition and conductivity improvement of tetragonal fast lithium ionic electrolyte $Li_7La_3Zr_2O_{12}$, *Solid State Ionics* 253(0) (2013) 137–142.

[156] Y. Wang, W. Lai: Phase transition in lithium garnet oxide ionic conductors $Li_7La_3Zr_2O_{12}$: The role of Ta substitution and H_2O/CO_2 exposure, *J. Power Sources* 275(0) (2015) 612–620.

[157] Y. Zhao, L. Daemen: Superionic conductivity in lithium-rich antiperovskites, *J. Am. Chem. Soc.* 134 (2012) 15042–15047.

[158] Y. S. Zhao, D. J. Weidner, J. B. Parise, D. E. Cox: Critical phenomena and phase transition of perovskite data for $NaMgF_3$ perovskite Part II, *Phys. Earth Planet. Inter.* 76 (1993) 17–34.

[159] M. O'keeffe, J.-O. Bovin: Solid electrolyte behavior of $NaMgF_3$: Geophysical implications, *Science* 206 (1979) 599–600.

[160] Y. Zhao: Crystal chemistry and phase transitions of perovskite in P-T-X space: data for $(K_xNa_{1-x})MgF_3$ perovskites, *J. Solid State Chem.* 141 (1998) 121–132.

[161] A. Yoshiasa, D. Sakamoto, H. Okudera, M. Sugahara, K. Ota, A. Nakatsuka: Electrical conductivities and conducting mechanisms of perovskite-type $Na_{1-x}K_xMgF_3$ (x = 0, 0.1, 1) and $KZnF_3$, *Z. Anorg. Allg. Chem.* 631 (2005) 502–506.

[162] G. Schwering, A. Hoennerscheid, L. Van Wuellen, M. Jansen: High lithium ionic conductivity in the lithium halide hydrates $Li_{3-n}(OH_n)Cl$ (0.83 < n < 2) and $Li_{3-n}(OH_n)Br$ (1 < n < 2) at ambient temperatures, *Chem. Phys. Chem.* 4 (2003) 343–348.

[163] J. Zhu, S. Li, Y. Zhang, J. W. Howard, X. Lu, Y. Li, Y. Wang, R. S. Kumar, L. Wang, Y. Zhao: Enhanced ionic conductivity with Li_7O_2Br ohse in Li_3OBr anti-perovskite solid electrolyte, *Appl. Phys. Lett.* 109 (2016) 101904.

[164] Z. Lu, C. Chen, Z. M. Baiyee, X. Chen, C. Niu, F. Ciucci: Defect chemistry and lithium transport in Li_3OCl anti-perovskite superionic conductors, *Phys. Chem. Chem. Phys.* 17 (2015) 32547–32555.

[165] D. J. Schroeder, A. A. Hubaud, J. T. Vaughey: Stability of the solid electrolyte Li_3OBr to common battery solvents, *Mater. Res. Bull.* 49 (2014) 614–617.

[166] S. Hull: Superionics: Crystal structures and conduction processes, *Rep. Prog. Phys.* 67 (2004) 1233–1314.

[167] S. Hull, D. A. Keen, P. A. Madden, M. Wilson: Ionic diffusion within the α^* and β phases of Ag_3SI, *J. Phys.: Condens. Matter* 19 (2007) 406214.

[168] A. Emly, E. Kioupakis, A. Van der Ven: Phase stability and transport mechanisms in anti-perovskite Li_3OCl and Li_3OBr superionic conductors, *Chem. Mater.* 25 (2013) 4663–4670.

[169] Y. Zhang, Y. S. Zhao, C. F. Chen: Ab-initio study of the stabilities of and mechanism of superionic transport in lithium-rich antiperovskites, *Phys. Rev. B* 87 (2013) 134303.

[170] G. W. Watson, S. C. Parker, A. Wall: Molecular dynamics simulation of fluoride-perovskites, *J. Phys.: Condens. Matter* 4 (1992) 2097–2108.

[171] R. Mouta, M. A. B. Melo, E. M. Diniz, C. W. A. Paschoal: Concentration of charge carriers, migration and stability in Li_3OCl solid electrolytes, *Chem. Mater.* 26 (2014) 7137–7144.

[172] M. H. Braga, J. A. Ferreira, V. Stockhausen, J. E. Oliveira, A. El-Azab: Novel Li_3ClO based glasses with superionic properties for lithium batteries, *J. Mater. Chem. A* 2 (2014) 5470–5480.

[173] Z. Deng, B. Radhakrishnan, S. P. Ong: Rational composition optimization of the lithium-rich $Li_3OCl_{1-x}Br_x$ anti-perovskite superionic conductors, *Chem. Mater.* 27 (2015) 3749–3755.

[174] K. Huang, R. S. Tichy, J. B. Goodenough: Superior perovskite oxide-ion conductor; strontium and magnesium doped $LaGaO_3$ I, Phase relationships and electrical properties, *J. Am. Ceram. Soc.* 81 (1998) 2565–2575.

[175] X. J. Lu, G. Wu, J. W. Howard, A. P. Chen, Y. S. Zhao, L. L. Daemen, Q. X. Jia: Li-rich anti-perovskite Li_3OCl films with enhanced ionic conductivity, *Chem.Commun.* 50 (2014) 11520–11522.

[176] H. Nguyen, S. Hy, E. Wu, Z. Deng, M. Samiee, T. Yersak, J. Luo, S. P. Ong, Y. S. Meng: Experimental and computational evaluation of a sodium-rich anti-perovskite for solid state electrolytes, *J. Electrochem. Soc.* 163(10) (2016) A2165–A2171.

[177] K. Takada, S. Kondo: Lithium ion conductive glass and its application to solid state batteries, *Ionics* 4 (1998) 42–47.

[178] N. Kamaya, K. Homma, Y. Yamakawa, M. Hirayama, R. Kanno, M. Yonemura, T. Kamiyama, Y. Kato, S. Hama, K. Kawamoto, A. Mitsui: A lithium superionic conductor, *Nat. Mater.* 10 (2011) 682–686.

[179] P. Bron, S. Johansson, K. Zick, J. Schmedt Auf Der Gunne, S. Dehnen, B. Roling: $Li_{10}SnP_2S_{12}$: An affordable lithium superionic conductor. *J. Am. Chem. Soc.* 135 (2013) 15694–15697.

[180] R. Kanno, T. Hata, Y. Kawamoto, M. Irie: Synthesis of a new lithium ionic conductor, thio-LISICON-lithium germanium sulfide system, *Solid State Ionics* 130 (2000) 97–104.

[181] R. Kanno, M. Maruyama: Lithium ionic conductor thio-LISICON: The Li_2S–GeS_2–P_2S_5 system, *J. Electrochem. Soc.* 148 (2001) A742–A746.

[182] S. Boulineau, M. Courty, J.-M. Tarascon, V. Viallet: Mechanochemical synthesis of Li-argyrodite Li_6PS_5X (X = Cl, Br, I) as sulfur-based solid electrolytes for all solid state batteries application, *Solid State Ionics* 221 (2012) 1–5.

[183] P. R. Rayavarapu, N. Sharma, V. K. Peterson, S. Adams: Variation in structure and Li^+-ion migration in argyrodite-type Li_6PS_5X (X = Cl, Br, I) solid electrolytes, *J. Solid State Electrochem.* 16 (2012) 1807–1813.

[184] R. P. Rao, S. Adams: Studies of lithium argyrodite solid electrolytes for all-solid-state batteries, *Phys. Status Solidi A* 208 (2011) 1804–1807.

[185] P. Knauth: Inorganic solid Li ion conductors: an overview, *Solid State Ionics* 180 (2009) 911–916.

[186] H. Y.-P. Hong: Crystal structure and ionic conductivity of $Li_{14}Zn(GeO_4)_4$ and other new Li^+ superionic conductor, *Mater. Res. Bull.* 13 (1978) 117–124.

[187] P. G. Bruce, A. R. West: Phase diagram of the LISICON, solid electrolyte system, Li_4GeO_4–Zn_2GeO_4, *Mater. Res. Bull.* 15 (1980) 379–385.

[188] M. Murayama, R. Kanno, M. Irie, S. Ito, T. Hata, N. Sonoyama, Y. Kawamoto: Synthesis of new lithium ionic conductor Thio-LISICON-Lithium silicon sulfides system, *J. Solid State Chem.* 168 (2002) 140–148.

[189] A. Kuhn, J. Kohler, B. V. Lotsch: Single-crystal X-ray structure analysis of the superionic conductor $Li_{10}GeP_2S_{12}$, *Phys. Chem. Chem. Phys.* 15 (2013) 11620–11622.

[190] Z. Q. Wang, M. S. Wu, G. Liu, X. L. Lei, B. Xu, C. Y. Ouyang: Elastic properties of new solid state electrolyte material Li_{10} GeP_2 S_{12}: A study from first-principles calculations, *Int. J. Electrochem. Sci.* 9 (2014) 562–568.

[191] S. P. Ong, Y. Mo, W. D. Richards, L. Miara, H. S. Lee, G. Ceder: Phase stability, electrochemical stability and ionic conductivity of the $Li_{10\pm1}MP_2X_{12}$(M = Ge, Si, Sn, Al or P, and X = O, S or Se) family of superionic conductors, *Energy Environ. Sci.* 6 (2013) 148–156.

[192] Y. Kato, R. Saito, M. Sakano, A. Mitsui, M. Hirayama, R. Kanno: Synthesis, structure and lithium ionic conductivity of solid solutions of $Li_{10}(Ge_{1-x}M_x)P_2S_{12}$ (M = Si, Sn), *J. Power Sources* 271 (2014) 60–64.

[193] J. M. Whiteley, J. H. Woo, E. Hu, K. W. Nam, S. H. Lee: Empowering the Lithium Metal Battery through a Silicon-Based Superionic Conductor, *J. Electrochem. Soc.* 161 (2014) A1812–A1817.

[194] A. Kuhn, O. Gerbig, C. Zhu, F. Falkenberg, J. Maier, B. V. Lotsch: A new ultrafast superionic Li-conductor: Ion dynamics in $Li_{11}Si_2PS_{12}$ and

comparison with other tetragonal LGPS-type electrolytes, *Phys. Chem. Chem. Phys.* 16 (2014) 14669–14674.

[195]   Y. Kato, S. Hori, T. Saito, K. Suzuki, M. Hirayama, A. Matsui, M. Yoneyama, H. Iba, R. Kanno: High-power all-solid-state batteries using sulfide superionic conductors, *Nature Energy* 1 (2016) 16030.

[196]   H.-J. Deiseroth, S.-T. Kong, H. Eckert, J. Vannahme, C. Reiner, T. Zaiß, M. Schlosser: Li_6PS_5X: A class of crystalline Li-rich solids with an unusually High Li^+ mobility, *Angew. Chem. Int. Ed.* 47(2008) 755–758.

[197]   R. B. Beeken, J. J. Garbe, J. M. Gillis, N. R. Petersen, B. W. Podoll, M. R. Stoneman: Electrical conductivities of the Ag_6PS_5X and the Cu_6PSe_5X (X = Br, I) argyrodites, *J. Phys. Chem. Solids* 66 (2005) 882–886.

[198]   M. S. Liao, W. H. E. Schwarz: Effective radii of the monovalent coin metals, *Acta Crystallogr. Sect. B* 50 (1994) 9–12.

Chapter 6

Amorphous/Glass and Glass-Ceramics Li-ion Conductive Ceramics

6.1 Amorphous/Glass Li-ion Conductor

The glassy ion conductor has demonstrated some advantages over crystalline ones, including isotropic Li-ion conduction, no grain-boundary resistance, ease of fabrication into thin-film form, wide range of composition, less sensitivity to ambient air and absence of transition metal ions that narrow the electrochemical window. In addition, the ionic conductivity in the amorphous glass is expected to be higher than that of the corresponding crystalline phase because of the open structure of the glass phase and the lower grain-boundary resistance. Figure 6.1 schematically shows the structure of SiO_2 crystal and SiO_2 glass [1]. The crystal possesses ordered structure in a long range, while the glass shows random and larger open structure.

   Li-ion conducting glasses can be divided into two categories, i.e. oxide and sulfide glasses [23]. For most of the oxide glasses, the Li-ion conductivity at room temperature is normally about 10^{-6}–$10^{-9}\,S\,cm^{-1}$ (Table 6.1) [6,24]. On the other hand, high Li-ion conductivity of 10^{-3}–$10^{-5}\,S\,cm^{-1}$ can be obtained for the sulfide glasses due to high polarization of sulfur ions [1,25,26]. Therefore, it is important to understand fundamentals of Li-ion conducting through

(a)

(b)

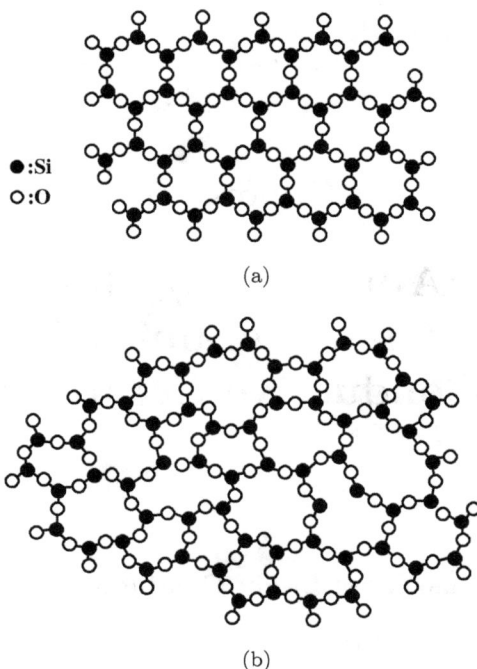

Fig. 6.1   Structure of (a) SiO_2 crystal and (b) SiO_2 glass.

the glass structure so that most suitable glassy Li-ion conductors can be designed and fabricated.

6.1.1   *Oxide glass*

Amorphous/glassy oxide conductor, denoted with the general formula of $Li_2O–M_xO_y$ ($M_xO_y = Al_2O_3, GeO_2, B_2O_3, P_2O_5, SiO_2$, etc.), [27–31] is normally formed by network-former oxides (e.g. B_2O_3, P_2O_5, SiO_2, etc.) and network-modifier oxides (Li_2O). In 1966, high Li-ion conduction of larger than $10^{-4}\,S\,cm^{-1}$ at about 350°C of $Li_2O–M_xO_y$ glass was found in glass composition of $Li_2O–SiO_2–B_2O_3$ [2]. It was found that the Li-ion conductivities could be detected when Li_2O was incorporated at more than 40 mol% in the glass. This finding proves that disordered structure can also support fast Li-ion conduction Lithium borate ($Li_2O–B_2O_3$), lithium

Table 6.1   Li-ion conductivity of amorphous solid electrolytes at room temperature.

Electrolyte	Conductivity ($S\,cm^{-1}$)	Reference
Oxide glass		
$50Li_2O-50B_2O_3$	7.1×10^{-8}	[2]
$31.8Li_2O-12.3LiCl-55.9B_2O_3$	3×10^{-6}	[3]
$50Li_2O-50SiO_2$	9.2×10^{-9}	[4]
$25Li_2O-25Al_2O_3-50SiO_2$	6.2×10^{-10}	[5]
$50Li_4SiO_4 \cdot 50Li_3BO_3$	5.2×10^{-2} (400 K)	[6]
$30Li_2SO_4 \cdot 45Li_2O \cdot 25P_2O_5$	4.2×10^{-7}	[7]
$Li_2O-Nb_2O_5$	1.6×10^{-6}	[8]
$Li_2O-Ta_2O_5$	5.0×10^{-6}	[8]
LiPON		
$Li_{2.9}PO_{3.3}N_{0.46}$	3.3×10^{-6}	[9]
$Li_{3.1}PO_{3.8}N_{0.16}$	2.0×10^{-6}	[9]
$Li_{3.3}PO_{3.8}N_{0.22}$	2.4×10^{-6}	[9]
PLD-deposited LiPON	1.6×10^{-6}	[10]
NIBAD[(1)]-LiPON	1.6×10^{-6}	[11]
EBRE[(2)]-LiPON	6.0×10^{-7}	[12]
Sulfide glass		
$30Li_2S-26B_2S_3-44LiI$	1.7×10^{-3}	[13]
$50Li_2S-17P_2S_5-33LiBH_4$	1.6×10^{-3}	[14]
$63Li_2S-36SiS_2-Li_3PO_4$	1.5×10^{-3}	[15]
$70Li_2S-30P_2S_5$	1.6×10^{-4}	[16]
$50Li_2S-50GeS_2$	4.0×10^{-5}	[17]
$45LiI-37Li_2S-18P_2S_5$	1.7×10^{-3}	[18]
$44LiI-30Li_2S-26B_2S_3$	1.7×10^{-3}	[13]
$60Li_2S-40SiS_2$	5.0×10^{-4}	[19]
$50Li_2S-50SiS_2$	1.2×10^{-4}	[20]
$60Li_2S-40SiS_2$	1.5×10^{-4}	[21]
$30LiCl-35Li_2S-35SiS_2$	2.7×10^{-4}	[20]
$0.01Li_3PO_4-63Li_2S-36SiS_2$	1.5×10^{-3}	[21]
$30LiI-42Li_2S-28SiS_2$	8.2×10^{-4}	[19]
$40LiI-36Li_2S-24SiS_2$	1.8×10^{-3}	[22]

Notes: (1) Nitrogen ion beam-assisted deposition. (2) E-beam reaction deposition.

silicate (Li_2O-SiO_2) and lithium phosphate ($Li_2O-P_2O_5$) glasses have been intensively investigated thus far. In these glasses, the Li-ion conductivity increases with Li_2O concentration [2, 3]. The same behavior is observed in other alkali metal ion conduction such as Na^+ and K^+. In the lithium borate and lithium silicate glasses, a sharp increase of Li-ion conductivity and decrease of activation energy can be achieved in the Li_2O concentration between 20 and 25 mol%, followed by much smaller changes with further increase of Li concentration [4, 32] However, a very high concentration of Li_2O (more than 50 mol% Li_2O) would cause devitrification. Addition of lithium halides and sulfate can avoid devitrification even at high Li concentration. Furthermore, the Li-ion conductivity of the glass can be significantly enhanced by addition of the lithium halides and sulfate compared with addition of Li_2O alone [3, 33–35]. For example, $31.8\,Li_2O–12.3\,LiCl–55.9\,B_2O_3$ glass revealed Li-ion conductivity of $3 \times 10^{-6}\,S\,cm^{-1}$ at 25°C, which was 50 times higher than that of $Li_2O–B_2O_3$ glass containing a similar amount of Li [3]. It is worth mentioning that the Li-ion conductivity of the glasses with the lithium halides increases with the size of halide ion. Table 6.2 summarizes the Li-ion conductivity of the lithium phosphate glass containing various lithium halides. It can be clearly seen that the order of the Li-ion conductivity of the phosphate glass was consistent with the order of the halide ion size, i.e. I > Br > Cl.

Analysis of the Raman spectra of these glasses showed almost no change in the vibrational modes of the P–O bonds, implying that the halogens were not incorporated in the polyphosphate chains. It was suggested that a relatively independent Li-halide substructure, which contributes to the enhanced alkali ion transport, is formed in the glass.

Another strategy to enhance Li-ion conductivity in the oxide glass is usage of two network-former oxides [24, 36–41]. For instance, the addition of another network-former oxide, SeO_2, into $Li_2O–B_2O_3$

Table 6.2   Li-ion conductivity of the lithium phosphate glasses containing various lithium halides [34].

Electrolyte	Conductivity ($S\,cm^{-1}$) at 25°C
$33LiI–67LiPO_3$	1.0×10^{-6}
$33LiBr–67LiPO_3$	3.2×10^{-7}
$30LiCl–70LiPO_3$	1.0×10^{-7}

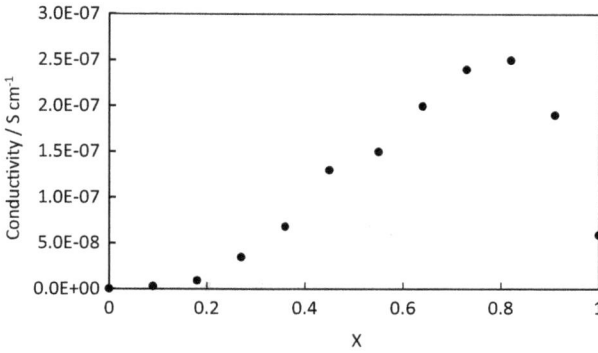

Fig. 6.2   Li-ion conductivity of $45\text{Li}_2\text{O}$–$55[x\text{B}_2\text{O}_3 - (1 - x)\text{P}_2\text{O}_5](0 \leq x \leq 1)$ glass.

glass can increase room-temperature ionic conductivity from 1.2×10^{-8} to $8 \times 10^{-7}\,\text{S}\,\text{cm}^{-1}$ [24]. Rauguenet *et al.* also systematically studied the effect of a network-former oxide on Li-ion conductivity using $45\text{Li}_2\text{O}$–$55[x\text{B}_2\text{O}_3 - (1-x)\text{P}_2\text{O}_5]$ $(0 \leq x \leq 1)$ glass (Fig. 6.2) [42]. Obviously, the usage of two network-former oxides (B_2O_5 and P_2O_5) enhances the conductivity due to the so-called "mixed anion effect" or "mixed glass former effect" [6,42]. The mixing anions are thought to make the degree of supercooling glasses smaller, which causes an enhancement of the Li-ion conductivity [6].

By utilizing rapid quenching techniques, some glasses, which normally cannot be prepared using normal sintering process, can be produced with Li-ion conductivity. For example, LiNbO_3 and LiTaO_3 glasses were prepared by rapid roll quenching. It should be noted that these glasses did not contain the network-former oxides [8]. Tatsumisago *et al.* developed a twin roller quenching system for preparation of glassy Li-ion conductor (Fig. 6.3) [43]. In this system, a sample that is put at one focal point of elliptical reactor is heated by a halogen lamp located at another focal point. After melting, droplets of the sample go into the twin roller and get cooled rapidly. The cooling rate of this system reached $\sim 10^9\,\text{K}\,\text{s}^{-1}$. This high cooling rate allows for the preparation of glasses with various compositions. A number of glassy conductors

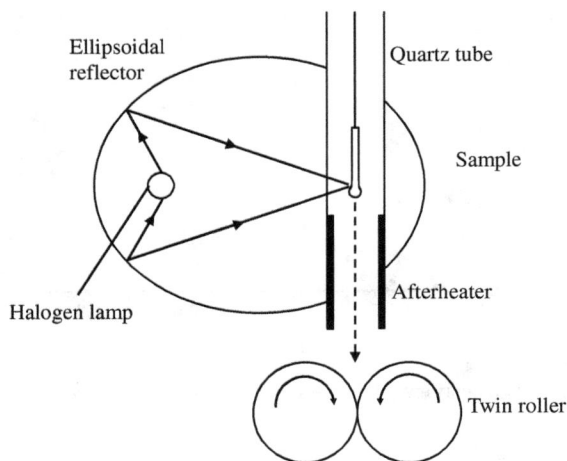

Fig. 6.3   Twin roller quenching system [43].

such as Li_3BO_3–Li_4GeO_4, Li_4SiO_4–Li_2WO_4, Li_3PO_4–Li_2SO_4 and Li_4SiO_4–Li_3BO_3 glasses are prepared.

6.1.2 *LiPON*

The most remarkable progress in amorphous/glassy oxide conductor was the development of $Li_xPO_yN_z$ (LiPON) in 1992, which was deposited by magnetron sputtering Li_3PO_4 target under a nitrogen flow condition [44]. Incorporation of nitrogen in the Li_2O–P_2O_5 network can effectively improve Li-ion conductivity, chemical stability against water and other physical properties such as increasing the glass transition temperature and hardness [45, 46]. The most important features of LiPON are its excellent chemical and electrochemical stability. LiPON exhibits high stability in contact with lithium metal and has a wide electrochemical window $(0 - 5.5\,V$ vs $Li^+/Li)$, which has made it a promising candidate for the solid electrolytes [47]. However, a recent study shows possible reaction between LiPON and metal lithium, confirmed by *in situ* XPS, but this reaction is thought to be limited [48]. The Li-ion conductivity of LiPON $(Li_{2.9}PO_{3.3}N_{0.46})$ is relatively low, $3.3 \times 10^{-6}\,S\,cm^{-1}$. However, this can be compensated by the very small thickness of LiPON [9]. In fact,

LiPON has been successfully used in thin film-type all-solid-state batteries [49].

In addition to magnetron sputtering preparation of LiPON [9], pulse laser deposition (PLD) [10], nitrogen beam-assisted deposition [11] and e-beam reaction evaporation [12] have been used for deposition of the LiPON thin films. Regardless of the preparation techniques, the Li ion conductivities of the thin films prepared by various methods have almost the same value, $\sim 10^{-6}\,\mathrm{S\,cm^{-1}}$. However, by applying bias to a substrate LiPON films with relatively high Li-ion conductivity could be obtained ($9.1 \times 10^{-6}\,\mathrm{S\,cm^{-1}}$ [50], $9.6 \times 10^{-6}\,\mathrm{S\,cm^{-1}}$ [51]).

XPS, Raman and IR spectroscopies have shown effective nitrogen incorporation in the Li_3PO_4 thin film using radio-frequency sputtering [52,53]. Successful incorporation of nitrogen into Li_3PO_4 thin film to form LiPON is dependent on numerous factors such as N_2 pressure, substrate composition, substrate temperature, deposition time and so on [54]. Although most studies have been performed by using LiPON thin film prepared in various experimental conditions, their reports are consistent with respect to nitrogen incorporation. The nitrogen incorporation consists of substitution of bridging oxygen with nitrogen followed by formation of PO_3N and PO_2N_2 anionic networks including doubly and triply coordinated nitrogen (Fig. 6.4) [55–57]. The formation of PO_3N and PO_2N_2 could be affected by Li/P ratio in lithium phosphate glass during deposition. Due to Li loss during the deposition process of Li_3PO_4 target, a mixture of orthophosphate and pyrophosphate units would be formed (Fig. 6.5) At Li/P=2 region, the oxynitride would contain $Li_2PO_2N_2$ moiety due to substitution of the bridging oxygen in the pyrophosphate unit (Fig. 6.6). At other Li/P ratios, the oxynitride is thought to contain $Li_7P_3O_8N_2$, $Li_5P_2O_6N$ and $Li_6P_2O_5N_2$ moieties depending on the local Li/P ratio [58].

The increase in conductivity due to nitrogen incorporation is obvious, but the mechanism is not still understood very clearly. It has been proposed that enhancement of conductivity was due to the decrease of electrostatic energy of LiPON with respect to Li_3PO_4 reference material [55] or an increase in the amount of triply

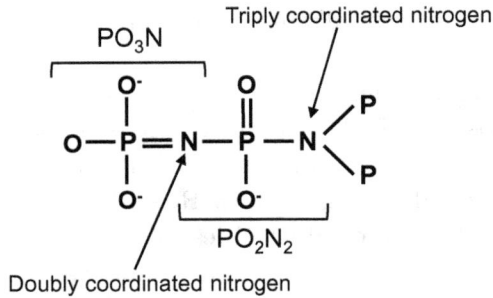

Fig. 6.4   PO$_3$N and PO$_2$N$_2$ networks.

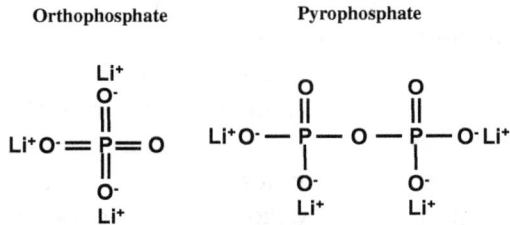

Fig. 6.5   Structures of orthophosphate and pyrophosphate.

coordinated nitrogen [59]. Muñoz *et al.* suggested that increase in the non-bridging oxygen content (Fig. 6.7), by the formation of P–N and P=N bonds which have higher covalent character as compared with P–O bonds, causes enhancement of the connectivity of the glass network, resulting in improvement of the ionic conductivity [60]. Mascaraque *et al.* also reported a decrease in the ratio of the bridging oxygen (BO) to non-bridging oxygen (NBO) by nitrogen incorporation directly linked the increase of ionic conductivity although the increase also depended on the lithium content [61]. The Li ion conductivity of LiPON thin film would be influenced by the lithium content, the ratio of BO/NBO, the nitrogen substitution and so on in a complex manner.

Apart from Li-ion conducting, LiPON also demonstrates some special optical properties [47]. For example, the thin film of LiPON (Li$_{3.13}$PO$_{1.69}$N$_{1.39}$) deposited on a silica glass has a visible-light

Li_2PO_2N

$Li_6P_2O_5N_2$

$Li_7P_3O_8N_2$

$Li_5P_2O_6N$

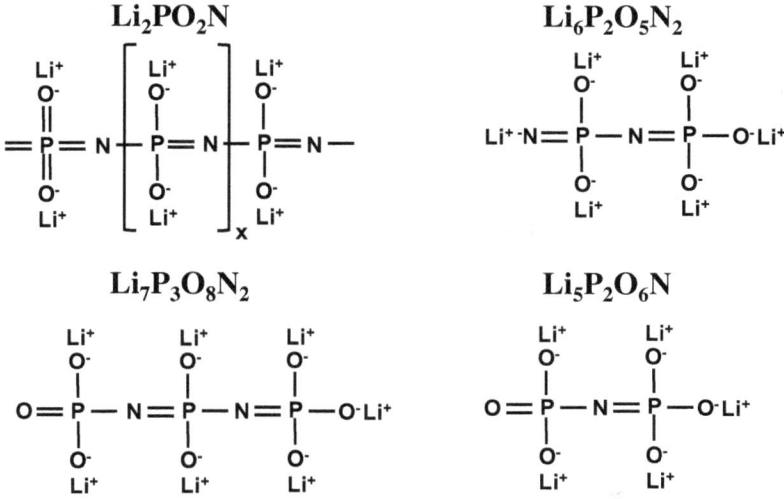

Fig. 6.6   Structures of various oxynitrides.

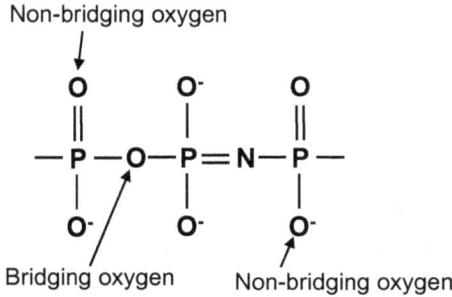

Non-bridging oxygen

Bridging oxygen   Non-bridging oxygen

Fig. 6.7   Bridging oxygen and non-bridging oxygen.

transparency higher than 80% [62], indicating that LiPON can also be used for electrochromic devices.

Due to the successful development of LiPON, i.e. enhancement of stability and conductivity by the substitution of a nitrogen for an oxygen atom, the same concept has been applied to other Li-ion conductive glasses. Li boron-oxynitride (LiBON) thin film is one of the examples that was prepared by radio-frequency sputtering under N_2 flow using $LiBO_2$ target like LiPON [63]. The Li-ion conductivity of $LiBO_{1.86}N_{0.09}$ thin film is $5 \times 10^{-8} \, S \, cm^{-1}$ at 26°C. The Li-ion conductivity increases with nitrogen content in the LiBON thin film, which could be controlled by a nitrogen flow rate during the sputtering [52]. Different from the LiPON containing doubly and triply coordinated nitrogen, only B–N–B unit in LiBON was observed. The nitrogen atoms also interact with Li atoms, which would contribute to the improvement of the conductivity [64]. In other examples, nitrogen-incorporated lithium borophosphate (LiBPON) and lithium silicophosphate (LiSiPON) also have been studied [65, 66].

6.1.3   *Sulfide glass*

Studies on ionic conduction in sulfides started from glasses [17]. The sulfide glasses tend to exhibit high ionic conductivity because the high polarization of sulfide ions weakens the interaction between lithium ions and anions. The high Li-ion conductivity of $\sim 10^{-3} \, S \, cm^{-1}$ in Li-containing sulfide glasses of $LiI–Li_2S–P_2S_5$ [18] and $LiI–Li_2S–B_2S_3$ [13] was discovered in the early 1980s. However, due to the high vapor pressure of P_2S_5 and B_2S_3 at high temperature, the synthesis of $LiI–Li_2S–P_2S_5$ and $LiI–Li_2S–B_2S_3$ glasses should be performed in a sealed container. The synthesis of sulfide glass under ambient pressure was achieved by using SiS_2 that has low vapor pressure. The Li-ion conductivity of $LiI–Li_2S–SiS_2$ system was comparable to that of $LiI–Li_2S–P_2S_5$ and $LiI–Li_2S–B_2S_3$ glasses [22]. Although the $LiI–Li_2S–SiS_2$ is unstable in contact with Li metal [67], Li_3PO_4 addition can dramatically improve the stability of $LiI–Li_2S–SiS_2$ glass [68]. Li_2SO_4 addition also provides the same effect. This oxysulfide glass with a general formula $Li_2S–SiS_2–Li_xMO_y$ (M=Si, P, Ge, B, Al etc.) has been given much attention because of its superior properties such as high conductivity, wide electrochemical window and high thermal stability against crystallization [69–71].

The mixed anion effect is also observed in the sulfide glass. The enhancement of Li-ion conductivity by the addition of lithium salts was reported due to increase in Li-ion concentration. The Li-ion conductivity of sulfide glasses was increased from an order of 10^{-4}

to $10^{-3}\,\mathrm{S\,cm^{-1}}$ at room temperature by the addition of lithium halides [13], lithium borohydride ($LiBH_4$) [14] and lithiumortho-oxosalts (Li_3PO_4) [15]. In general, it becomes more difficult to achieve glass phase with increase in Li-ion concentration because of the ease of crystallization during the cooling process. The rapid quenching technique is helpful to avoid the crystallization and allows for the preparation of the sulfide glass with wider composition range. Aotani *et al.* prepared Li_3PO_4–Li_2S–SiS_2 glass using a twin roller quenching technique [15]. The Li-ion conductivity of the obtained glass was $1.5 \times 10^{-3}\,\mathrm{S\,cm^{-1}}$, which was higher than that prepared by liquid N_2 quenching. Additionally, the sulfide glasses can be simply prepared by mechanical ball-milling [21]. In this case, even a heating process is not necessary. This is beneficial in mass production of the sulfide glasses and is another advantage when compared with the oxide glasses.

The major shortcoming of sulfide glasses is that they must be handled in an inert atmosphere because the sulfide glasses can react with ambient moisture [72]. When the sulfide glasses are exposed to moisture in air, hydrolysis of the glasses proceeds, resulting in generation of toxic H_2S gas. The generation of H_2S depends on the glass composition. In Li_2S–P_2S_5 glass, the amount of H_2S decreased with increase of Li_2S content and reached the minimum value ($0.01\,\mathrm{cm^3\,g^{-1}}$) at $75\,\mathrm{mol\%}$ Li_2S. Then, the amount increased again [73]. Muramatsu *et al.* investigated the structural change of Li_2S–P_2S_5 glass in air [74] and noted that $P_2S_7^{4-}$ ion which is the main structural unit of $67Li_2S$–$33P_2S_5$ glass [75], decomposed to form $-SH$ and $-OH$ groups by hydrolysis in moisture. The $-SH$ group further reacted with H_2O and finally H_2S and two PS_3OH (Fig. 6.8(a)) Moreover, Li_2S reacted with H_2O and then formed $LiSH$ and $LiOH$. The $LiSH$ was hydrolyzed to form $LiOH$ and H_2S (Fig. 6.8(b)).

On the contrary $75Li_2S$–$25P_2S_5$ glass composed of PS_4^{3-} ion does not show obvious structural change after exposure to air for 1 day. Therefore, it is concluded that this high stability of PS_4^{3-} ion in air is associated with excellent high stability of $75Li_2S$–$25P_2S_5$ glass. Furthermore, $75Li_2S$–$25P_2S_5$ glass does not generate H_2S under O_2 or N_2 gas with low humidity, which implies that the sulfide glass

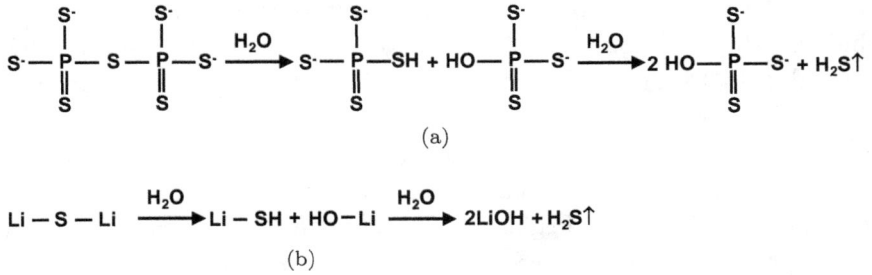

(a)

(b)

Fig. 6.8 H_2S generation from $P_2S_7^{4-}$ and Li_2S.

would be stable in dry air. Partial substitution of oxides (Li_2O or P_2O_5) for sulfides (Li_2S or P_2S_5) can be effective in suppressing the H_2S generation. The amount of H_2S generation in Li_2O-added $75Li_2S-25P_2S_5$ glass is lower than that in $75Li_2S-25P_2S_5$ glass [76]. Moreover, partial replacement of P_2S_5 with P_2O_5 demonstrated reduction of H_2S generation. Addition of metal oxides was also reported to be an effective approach to suppress H_2S generation. Metal oxides such as Fe_2O_3, ZnO and Bi_2O_3 are able to absorb or adsorb H_2S through acid–base reaction. A composite with 90 mol% of $75Li_2S-21P_2S_5-4P_2O_5$ glass and 10 mol% ZnO prepared by mechanical milling decreased the rate of H_2S generation when exposed air [77]. However, the Li-ion conductivity of the sulfide glasses tends to be decreased by the addition of oxides.

6.2   Glass-Ceramic Li-ion Conductor

Glass-ceramic conductors can be defined as partially crystallized glass (amorphous) ceramics. In other words, the crystal grains of the ceramic are surrounded by its precursor glass. Precipitation of thermodynamically stable crystalline phases from the precursor glass is useful for reducing grain-boundary resistance. Since grain-boundaries among crystal domains are filled with amorphous phases, it leads to high conductivity (Fig. 6.9, Table 6.3).

Therefore, the conductivity of glass-ceramics is generally higher than that of their ceramics counterpart. NASICON-type glass-ceramics can be obtained in $Li_2O-Al_2O_3-TiO_2-P_2O_5$ (LATP) [78]

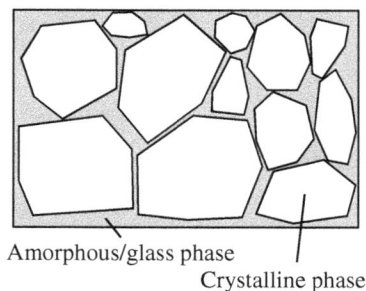

Amorphous/glass phase

Crystalline phase

Fig. 6.9   Schematic illustration of glass-ceramics.

Table 6.3   Li-ion conductivity of glass-ceramic electrolytes at room temperature.

Electrolyte	Conductivity ($S\,cm^{-1}$)	Reference
Oxide glass-ceramics		
LATP	1.3×10^{-3}	[78]
Microwave-LATP	5.33×10^{-4}	[79]
$Li_2O–TiO_2–P_2O_5 - SiO_2$	7×10^{-4}	[80]
LAGP	4×10^{-4}	[81]
Li_2O–add LAGP	7.25×10^{-4}	[82]
B_2O_3-added LAGP	6.9×10^{-4}	[83]
Sulfide glass-ceramics		
$70Li_2S–30P_2S_5$	3.2×10^{-3}	[84]
$78Li_2S–22P_2S_5$	1.1×10^{-3}	[85]
$Li_2S–P_2S_5–P_2S_3$	5.4×10^{-3}	[86]
$Li_2S–P_2S_5–LiI$	2.7×10^{-3}	[87]
$70Li_2S–30P_2S_5$	1.7×10^{-2}	[88]

and $Li_2O–Al_2O_3–GeO_2–P_2O_5$ (LAGP) [89] systems. Usually, these glass-ceramics are prepared by quenching a melt of raw material heated at $>1300°C$, followed by crystallization process. The transparent glass changed into a milky glass-ceramic composite after the crystallization (Fig. 6.10).

The properties of the glass-ceramics are largely influenced by the crystallization process, such as Li-ion conductivity of $Li_{1.5}Al_{0.5}$ $Ge_{1.5}(PO_4)_3$ (LAGP) glass-ceramics [90]. Within a wide range of crystallization temperatures, the TG-DTA curve of LAGP glass

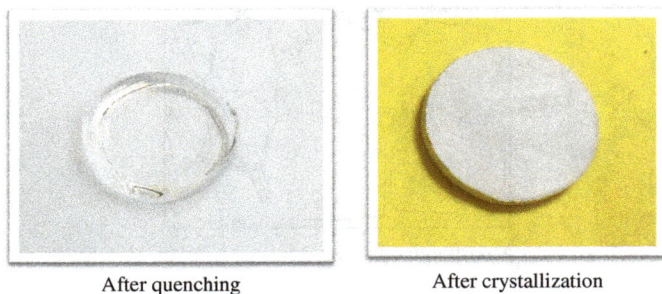

After quenching   After crystallization

Fig. 6.10   Glass-ceramics after quenching and post-crystallization.

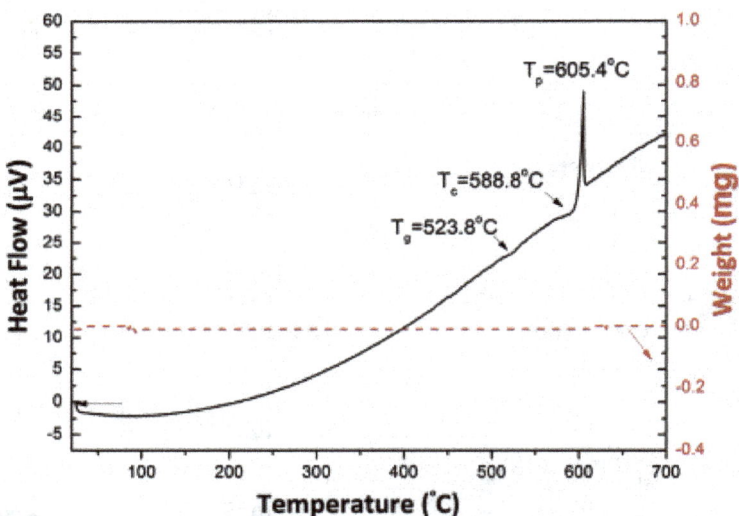

Fig. 6.11   TG-DTA curve of LAGP glass.

(Fig. 6.11) shows no weight change, while a strong exothermic peak appeares at 605.4°C, which can be attributed to crystallization of the LAGP glass. Based on this exothermic peak, crystallization temperature can be determined. Since crystallization is not only thermodymic dependent but also kinetic dependent, crystallinity of the glass-ceramics increases with prolonged high-temperature annealing. Figure 6.12 reveals XRD patterns of the LAGP glass after crystallization at various temperatures for 8 h. In all samples, most

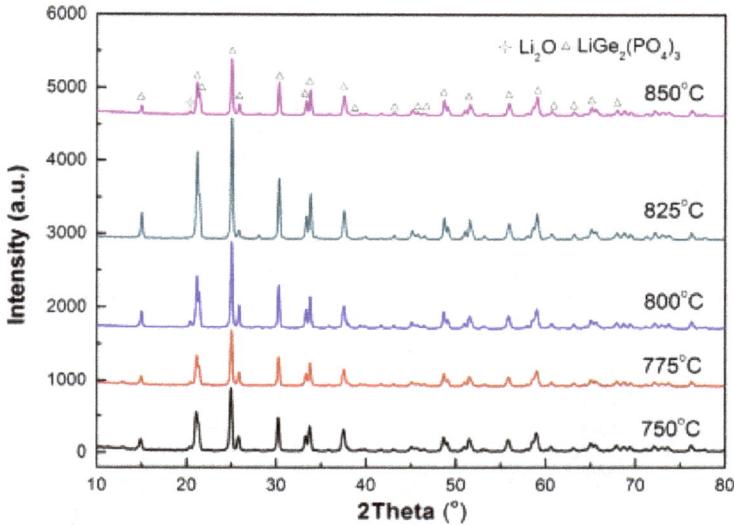

Fig. 6.12   XRD patterns of LAGP glass after subjecting to various crystallization temperatures for 8 h.

of the peaks could be assigned to $LiGe_2(PO_4)_3$ with rhombohedral structure, indicating that LAGP is the dominant phase in all samples. However, Li_2O is also observed in all samples, and the amount of Li_2O slightly increased with the crystallization temperature. This is caused by a phase separation. The crystllization temperature influences not only crystallinity of the glass-ceramics, but also impurity formation. Table 6.4 shows the lattice parameters of the LAGP after crystallization. The lattice parameters were the same among all samples, indicating that the crystallization temperature does not affect the lattice parameters of LAGP crystal.

As expected, the crystallization temperature affects crystal grain size and morphology of the glass-ceramics. Figure 6.13 displays SEM images of pristine LAGP glass and glass after crystallization at various temperatures for 8 h. The LAGP glass had a feature-less morphology, indicating its amorphous nature (Fig. 6.13(a)). The crystal grains were observed after crystallization at 750°C (Fig. 6.13(b)), while the edges of grains were not clear because the glass component existed among the grains. With increase of

Table 6.4 Lattice parameters of LAGP crystallized at various temperatures.

Crystallized temperatures (°C)	a (Å)(±0.002)	c (Å)(±0.002)	V (Å³)(±0.04)
750	8.263	20.654	1221.2
775	8.262	20.654	1221.1
800	8.262	20.656	1221.0
825	8.262	20.653	1221.0
800	8.262	20.639	1220.1

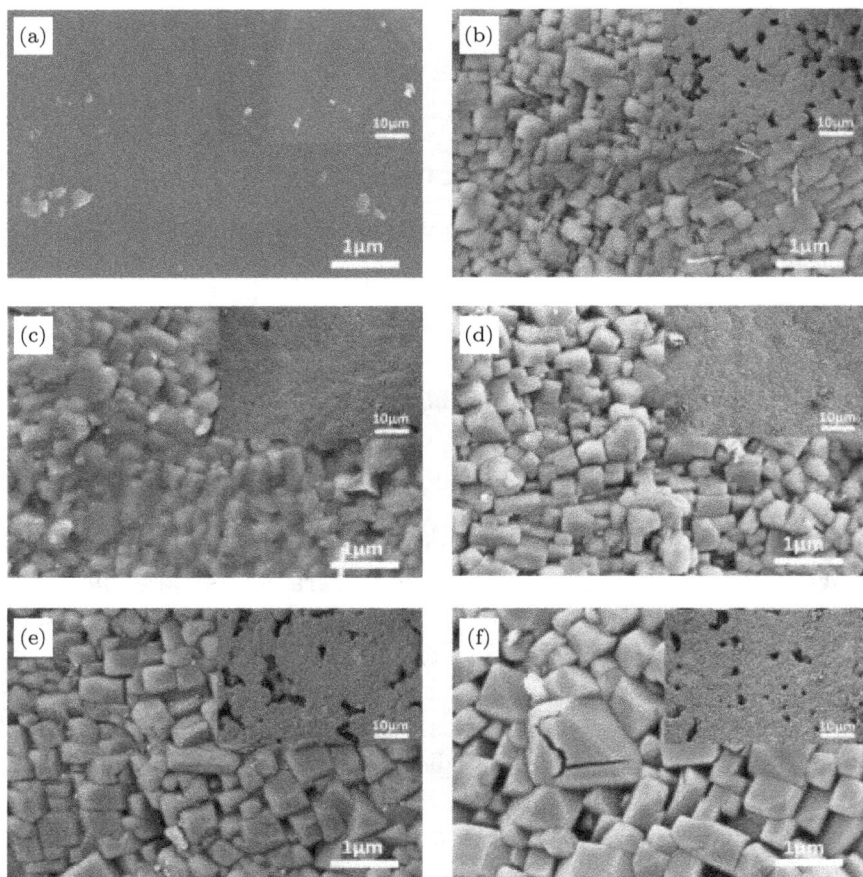

Fig. 6.13 SEM micrographs of (a) pristine LAGP glass and the glass after crystallization at (b) 750°C, (c) 775°C, (d) 800°C, (e) 825 and (f) 8°C.

the crystallization temperature, the grain size of the crystal phase became larger, whereas the lattice parameter remained unchanged (Table 6.4). The low-magnification micrographs (shown in the inset) show reversal formation of pores at low crystallization temperature. The pores become progressively less in number with increase of the crystallization temperature, and finally disappear after crystallization at 800°C (Fig. 6.13(d)). The pores appear again with further increase of the crystallization temperature. The sample crystallized at 800°C shows the most compact structure and relatively large grains.

Figure 6.14 reveals Li-ion conductivity of LAGP glass-ceramics crystallized at various temperatures. The conductivity increases rapidly with crystallization temperature up to 800°C and then decreases with further increase in the temperature. Increase in total conductivity is attributed to increase in grain-boundary and bulk conductivity. This good balance of bulk and grain-boundary conductivities would be one reason for the highest conductivity observed in the sample crystallized at 800°C.

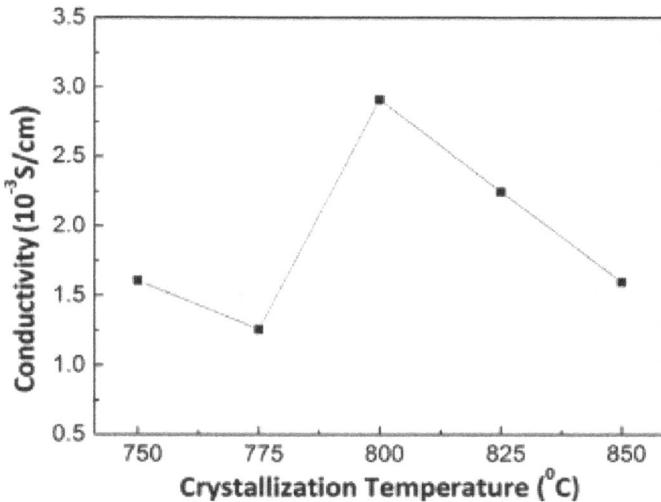

Fig. 6.14 Li-ion conductivity of LAGP glass-ceramic crystallized at various temperatures.

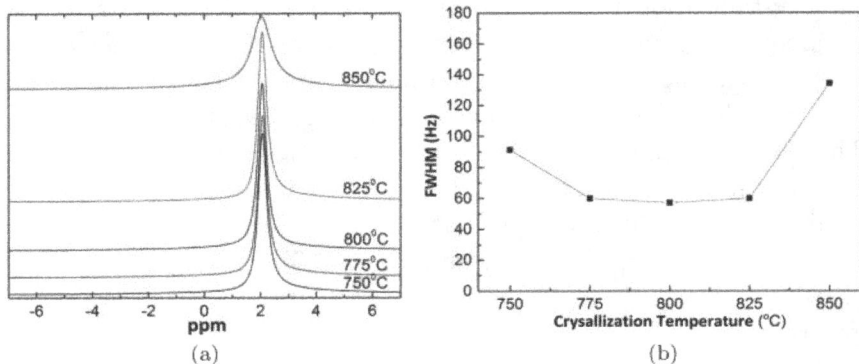

(a)   (b)

Fig. 6.15 ^{7}Li MAS NMR spectra of LAGP glass after crystallization at various temperatures and (b) FWHM obtained from the NMR spectra.

Additionally, it was reported that the crystallization temperature also influenced Li-ion mobility. Figure 6.15 reveals ^{7}Li MAS NMR spectra of LAGP glass after crystallization at various temperatures. There is a peak observed at around 2 ppm independent of the crystallization temperature. The full width at half maximum (FWHM) of the peak relates to the mobility of Li-ion in the crystal grains. The FWHM is shown in Fig. 6.15(b). A low FWHM value was observed from 750°C to 825°C, indicating increase of Li-ion mobility in this temperature range. The crystallization temperature strongly influences crystal and grain-boundary structures and the mobility of Li-ion of LAGP that determine Li-ion conductivity. Thus, the crystallization temperature should be chosen carefully.

Since crystallinity is crystallization duration dependent, duration of crystallization also influences the conductivity of LAGP glass-ceramics. Figure 6.16 displays SEM images of LAGP glass developed after different crystallization duration. The influence of crystallization duration is similar to that of crystallization temperature. The growth of crystal grains is observed with increase of crystallization duration. As shown in Fig. 6.16, formation of pores is observed after 4 hours of crystallization; the pores disappear after 8 h crystallization, and then form again with longer crystallization duration. The highest Li-ion conductivity is obtained after 8 h of crystallization

Fig. 6.16 SEM images of LAGP glass after crystallization at 800°C for different crystallization duration.

Table 6.5 Li-ion conductivity of LAGP glass after crystallization at 800°C in different crystallization time.

Crystallization time (h)	4	8	12
Li-ion conductivity ($\times 10^{-3}\,\mathrm{S\,cm^{-1}}$)	0.66	2.91	0.64

(Table 6.5) where the most compact morphology and relatively large crystal grains are achieved. This feature is consistent with LAGP sample with the highest Li-ion conductivity obtained in a series of crystallization temperature experiment.

Similar studies were performed in LATP glass-ceramics. Soman *et al.* studied microstructural and ionic conductivity evolution of LATP glass-ceramics crystallized at 700°C for 0–60 h [91]. The grain size increased with prolonging the crystallization duration. However, the ionic conductivity abruptly increased after crystallization for a short time of 0.5 h. It was concluded that the sudden increase of conductivity occurred at a crystalline volume content of more than 30% and this was consistent with the percolation law. Homogeneous nucleation of crystal is also important in the crystallization process as pointed out by Narváez-Semanate *et al.* [92]. To study the importance of nucleation, LATP is first nucleated at a temperature 10°C lower than glass transition temperature before further crystallization, which is called the crystal growth step. The ionic conductivity of LATP glass-ceramics with the nucleation process is found to be higher than that without the nucleation process. As for heating method, the benefit of microwave heating was reported [79]. By microwave heating, large crystal grain of LATP was obtained. The Li-ion conductivity of LATP glass-ceramics with microwave heating was five times higher than that with conventional heating. Interaction between electromagnetic radiation and the sample and rapid heating would cause the enhancement of conductivity.

Partial substitution of Al, Ti (Ge) and/or P by heteroatoms and addition of other ceramics is also an effective approach to improve ionic conductivity of the glass-ceramics. It has been shown that the conductivity of LATP glass-ceramics increases by partial substitution of SiO_2 for P_2O_5 [93]. The conductivity improvement by SiO_2 substitution was also observed in Al-free samples. The Li-ion conductivity of the Li_2O–TiO_2–P_2O_5 glass-ceramics increases to $7 \times 10^{-4}\,S\,cm^{-1}$ by SiO_2 substitution [80]. This is more than 100 times higher than that of SiO_2-free sample [94]. The increase of ionic conductivity of the LAGP glass-ceramics has been also achieved by excess Li_2O addition [82]. The excess Li_2O is not only incorporated into the crystal lattice of LAGP but also exists as a secondary phase acting as a nucleating agent during crystallization of LAGP glass. This nucleation causes improvement in the connectivity between glass and ceramic grains, resulting in increase of conductivity. The same effect was also reported on B_2O_3 addition to LAGP [83]. The nonlinearity in Arrhenius total conductivity plots was observed in B_2O_3–LAGP system. This is considered to be due to the space charge effect of $AlPO_4$ impurity phase [95]. Figure 6.17 shows the Arrhenius plot of $AlPO_4$-containing LAGP glass-ceramics. The Arrhenius plot comprised two linear lines, which intersect at 47°C. This can be attributed to the formation of a $AlPO_4$:Li^+ complex which causes the space charge effect. $AlPO_4$ forms the $AlPO_4$:Li^+ complex, and the complex becomes a source of space charge that mediates ion transport. The complex is stable up to 25–45°C. Accordingly, the space charge effect disappears at high temperature. This would be a reason for the nonlinearity of the Arrhenius plot. This is a

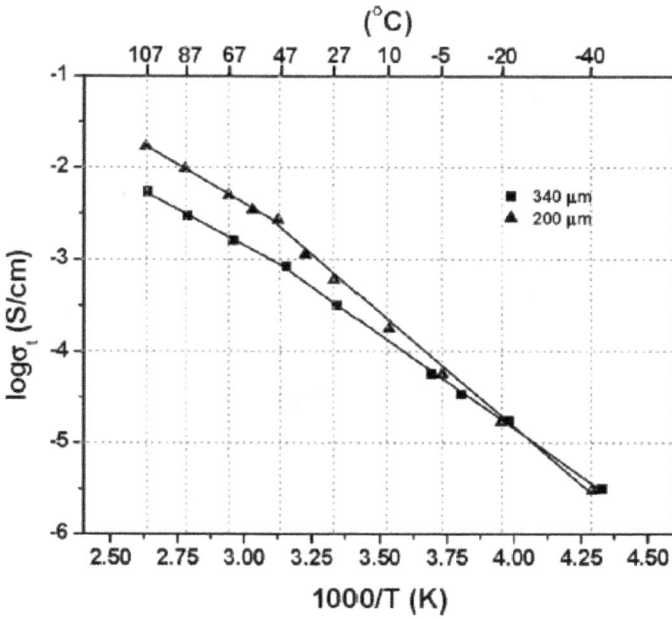

Fig. 6.17   Arrhenius plot of AlPO$_4$-containing LAGP glass-ceramics.

good example that the formation of impurity phase does not always provide negative effect.

On the other hand, sulfide glass-ceramics have not been intensively studied like the oxide glass-ceramics. The grain-boundary resistance in the crystalline sulfide ceramics can be reduced easily by simple cold pressing due to the elastic nature of the crystalline sulfide ceramics. Therefore, the sulfide glass-ceramics have not been given much attention. Crystallization of the sulfide glass usually takes place between 200°C and 400°C, which is much lower than that of oxide glass-ceramics (usually 700–1000°C). The reported conductivity of sulfide glass-ceramics is about 10^{-3} S cm^{-1} [84, 85, 96]. For example, 70Li$_2$S–30P$_2$S$_5$ glass possesses the highest Li-ion conductivity of 3.2×10^{-3} S cm^{-1} after crystallization at 360°C [97]. The Li-ion conductivity is strongly affected by the crystallization temperature similar to the oxide glasses. The conductivity of the 70Li$_2$S–30P$_2$S$_5$ glass dropped to 1.1×10^{-6} S cm^{-1} after crystallization at 550°C

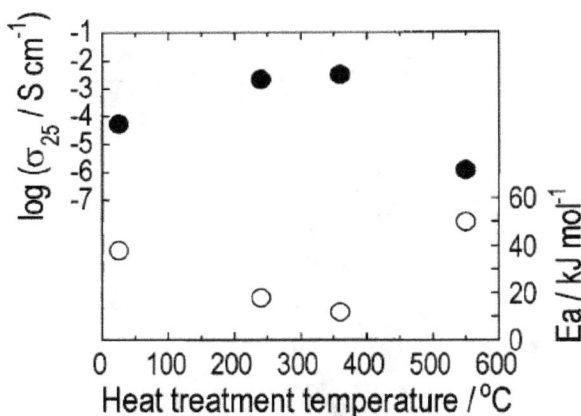

Fig. 6.18   Relationship between Li-ion conductivity and crystallization temperature of $70Li_2S-30P_2S_5$ glass.

(Fig. 6.18). The conductivity could be improved by LiI [87] or P_2S_3 [86] addition. Furthermore, extremely high conductivity of $1.7 \times 10^{-2}\,S\,cm^{-1}$ of $Li_2S-P_2S_5$ glass-ceramics can be obtained by densification of the glass through hot pressing of glass powder at $300°C$ [88]. In this process, crystallization and densification are achieved simultaneously. The conductivity is even higher than that of $Li_{10}GeP_2S_{12}$ crystal. The hot-pressed glass-ceramic pellets show negligible grain-boundary resistance, so that remarkably high total conductivity can be achieved. Moreover, the hot-pressed $Li_2S-P_2S_5$ glass-ceramic pellets show reasonable stability when in contact with Li metal.

References

[1] M. Tatsumisago: Glassy materials based on Li_2S for all-solid-state lithium secondary batteries, *Solid State Ionics* 175 (2004) 13–18.

[2] K. Otto: Electrical conductivity of $SiO_2-B_2O_3$ glasses containing lithium or sodium, *Phys. Chem. Glasses* 7(1) (1966) 29–37.

[3] A. Levasseur, J. C. Brethous, J. M. Reau, P. Hagenmuller: Comparative-study of ionic-conductivity of lithium in vitreous halogen borate, *Mater. Res. Bull.* 14(7) (1979) 921–927.

[4] R. J. Charles: Metastable liquid immiscibility in alkali metal oxide-silica systems, *J. Am. Ceram. Soc.* 49 (1966) 55–62.

[5] R. M. Biefeld, R. T. Johnson, R. J. Baughman: Effects of composition changes, substitutions, and hydrostatic pressure on the ionic conductivity in lithium aluminosilicate and related beta-eucryptite materials, *J. Electrochem. Soc.* 125 (1978) 179–185.

[6] M. Tatsumisago, N. Hachida, T. Minami: Mixed anion effect in conductivity of rapidly quenched Li_4SiO_4–Li_3BO_3 glasses, *YogyoKyokaishi* 95 (1987) 197–201.

[7] M. Ganguli, M. H. Bhat, K. J. Rao: Lithium ion transport in Li_2SO_4–Li_2O–P_2O_5 glasses, *Solid State Ionics* 122 (1999) 23–33.

[8] A. M. Glass, K. Nassau, T. J. Negran: Ionic conductivity of quenched alkali niobate and tantalate glasses, *J. Appl. Phys.* 49 (1978) 4808–4811.

[9] J. B. Bates, N. J. Dudney, G. R. Gruzalski, R. A. Zuhr, A. Choudhury, C. F. Luck: Fabrication and characterization of amorphous lithium electrolyte thin films and rechargeable thin-film batteries, *J. Power Sources* 43(1–3) (1993) 103–110.

[10] S. ZHaon, Z. Fu, Q. Qin: A solid-state electrolyte lithium phosphorus oxynitride film prepared by pulsed laser deposition, *Thin Solid Films* 415 (2002) 108–113.

[11] F. Vereda, R. B. Goldner, T. E. Haas, P. Zerigian: Rapidly grown IBAD LiPON films with high li-ion conductivity and electrochemical stability, *Electrochem. Solid State Lett.* 5 (2002) A239–A241.

[12] W.-Y. Liu, Z.-W. Fu, C.-L. Li, Q.-Z. Qin: Lithium phosphorus oxynitride thin film fabricated by a nitrogen plasma-assisted deposition of e-beam reaction evaporation, *Electrochem. Solid State Lett.* 7 (2004) J36–J39.

[13] H. Wada, M. Menetrier, A. Levasseur, P. Hagenmuller: Preparation and ionic conductivity of new B_2S_3–Li_2S–LiI glasses, *Mater. Res. Bull.* 18 (1983) 189–193.

[14] A. Yamauchi, A. Sakuda, A. Hayashi, M. Tatsumisago: Preparation and ionic conductivities of $(100 - x)(0.75Li_2S \cdot 0.25P_2S_5) \cdot xLiBH_4$ glass electrolytes, *J. Power Sources* 244 (2013) 707–710.

[15] N. Aotani, K. Iwamoto, K. Takada, S. Kondo: Synthesis and electrochemical properties of lithium ion conductive glass, Li_3PO_4–Li_2S–SiS_2, *Solid State Ionics* 68 (1994) 35–39.

[16] Z. Zhang, J .H. Kennedy: Synthesis and characterization of the B_2S_3–Li_2S, the $P_2S5 - Li_2S$ and the B_2S_3–P_2S_5–Li_2S glass systems, *Solid State Ionics* 38 (1990) 217–224.

[17] M. Ribes, B. Barrau, J. L. Souquet : Sulfide glasses: Glass forming region, structure and ionic conduction of glass in Na_2S–$XS_2(X = Si, Ge)$, Na_2S–P_2S_5 and Li_2S–GeS_2 systems, *J. Non-Cryst Solids* 38 (1980) 271–276.

[18] R. Mercier, J.-P. Malugani, B. Fahys, G. Robert: Superionic conduction in Li_2S–P_2S_5–LiI-glasses, *Solid State Ionics* 5 (1981) 663–666.

[19] A. Pradel, M. Ribes: Electrical properties of lithium conductive silicon sulfide glasses prepared by twin roller quenching, *Solid State Ionics* 18/19 (1986) 351–355.

[20] J. H. Kennedy, S. Sahami, S. W. Shea, Z. Zhang: Preparation and conductivity measurements of SiS_2–Li_2S glasses doped with LiBr and LiCl, *Solid State Ionics* 18/19 (1986) 368–371.

[21] H. Morimoto, H. Yamashita, M. Tatsumisago, T. Minami: Mechanochemical synthesis of new amorphous materials of $60Li_2S \cdot 40SiS_2$ with high lithium ion conductivity, *J. Am. Ceram. Soc.* 82 (1999) 1352–1354.

[22] J. H. Kennedy, Y. Yang: A highly conductive Li^+-glass system: $(1-x) \cdot (0.4SiS_2–0.6Li_2S) - xLiI$, *J. Electrochem. Soc.* 133 (1986) 2437–2438.

[23] C. Cao, Z.-B. Li, X.-L. Wang, X.-B. Zhao, W.-Q. Han: Recent advances in inorganic solid electrolytes for lithium batteries, *Front. Energy Res.* 2 (2014) 1–10.

[24] C. H. Lee, K. H. Joo, J. H. Kim, S. G. Woo, H. J. Sohn, T. Kang, Y. Park, J. Y. Oh: Characterizations of a new lithium ion conducting Li_2O–SeO_2–B_2O_3 glass electrolyte, *Solid State Ionics* 149 (2002) 59–65.

[25] N. Machida, T. Shigematsu: An all-solid-state lithium battery with sulfur as positive electrode materials, *Chem. Lett.* 33 (2004) 376–377.

[26] T. Ohtomo, A. Hayashi, M. Tatsumisago, Y. Tsuchida, S. Hama, K. Kawamoto: All-solid-state lithium secondary batteries using the $75Li_2S \cdot 25P_2S_5$ glass and the $70Li_2S$–$30P_2S_5$ glass-ceramic as solid electrolyte, *J. Power Sources* 233 (2013) 231–235.

[27] R. J. Charles: Some structural and electrical properties of lithium silicate glasses, *J. Am. Ceram. Soc.* 46(5) (1963) 235–238.

[28] S. W. Martin: Ionic conduction in phosphate glasses, *J. Am. Ceram. Soc.* 74(8) (1991) 1767–1784.

[29] M. M. A. Levasseur, B. Cales, J.-M. Reau, P. Hagenmuller: Ionic conductivity of lithium in glasses of B_2O_3–Li_2O–LiCl system, *Mater. Res. Bull.* 13(3) (1978) 205–209.

[30] M. Doreau, A. Abouelanouar, G. Robert: vitreous regions, structure and electrical-conductivity of glass in the LiCl–Li_2O–P_2O_5 system, *Mater. Res. Bull.* 15(2) (1980) 285–294.

[31] M. K. Murthy, J. Ip: Studies in germanium oxide systems: I, Phase equilibria in the system Li_2O–GeO_2, *J. Am. Ceram. Soc.* 47(7) (1964) 328–331.

[32] R. M. Hakim, D. R. Uhlmann: Electrical conductivity of alkali silicate glasses, *Phys. Chem. Glasses* 12 (1971) 132–138.

[33] K. Tanaka, T. Yoko, H. Yamada, K. Kamiya: Structure and ionic conductivity of LiCl–Li_2O–TeO_2 glasses, *J. Non-Cryst. Solids* 103(2–3) (1988) 250–256.

[34] J. P. Malugani, G. Robert: Ion conductivity in $LiPO_3$–LiX glasses (X=I,Br,Cl), *Mater. Res. Bull.* 14(8) (1979) 1075–1081.

[35] K. Tanaka, T. Yoko, K. Kamiya, H. Yamada, S. Sakka: Properties of oxybromide tellurite glasses in the system LiBr–Li_2O–TeO_2, *J. Non-Cryst. Solids* 135(2–3) (1991) 211–218.

[36] T. Tsuchiya, T. Moriya: Anomalous behavior of physical and electrical properties in borophosphate glasses containing R_2O and V_2O_5, *J. Non-Cryst. Solids* 38–39(Part 1) (1980) 323–328.

[37] A. Magistris, G. Chiodelli, M. Villa: Lithium borophosphate vitreous electrolytes, *J. Power Sources* 14(1–3) (1985) 87–91.

[38] B. V. R. Chowdari, K. Radhakrishnan: Electrical and electrochemical characterization of $Li_2O:P_2O_5:Nb_2O_5$-based solid electrolytes, *J. Non-Cryst. Solids* 110(1) (1989) 101–110.

[39] B. V. R. Chowdari, K. Radhakrishnan: Ionic conductivity studies of the vitreous $Li_2O–P_2O_5–Ta_2O_5$ system, *J. Non-Cryst. Solids* 108(3) (1989) 323–332.

[40] Y.-I. Lee, J.-H. Lee, S.-H. Hong, Y. Park: Li-ion conductivity in $Li_2O–B_2O_3–V_2O_5$ glass system, *Solid State Ionics* 175(1–4) (2004) 687–690.

[41] C. E. Kim, H. C. Hwang, M. Y. Yoon, B. H. Choi, H. J. Hwang: Fabrication of a high lithium ion conducting lithium borosilicate glass, *J. Non-Cryst. Solids* 357(15) (2011) 2863–2867.

[42] B. Raguenet, G. Tricot, G. Silly, M. Ribes, A. Pradel: The mixed glass former effect in twin-roller quenched lithium borophosphate glasses, *Solid State Ionics* 208 (2012) 25–30.

[43] M. Tatsumisago, T. Minami, M. Tanaka: Rapid quenching technique using thermal-image furnace for glass preparation, *J. Am. Ceram. Soc.* 64 (1981) C97–C98.

[44] J. B. Bates, N. J. Dudney, G. R. Gruzalski, R. A. Zuhr, A. Choudhury, C. F. Luck, J. D. Robertson: Electrical properties of amorphous lithium electrolyte thin films, *Solid State Ionics* 53–56 (1992) 647–654.

[45] R. Marchand: Nitrogen-containing phosphate glasses, *J. Non-Cryst. Solids* 56(1–3) (1983) 173–178.

[46] R. W. Larson, D. E. Day: Preparation and characterization of lithium phosphorus oxynitride glass, *J. Non-Cryst. Solids* 88(1) (1986) 97–113.

[47] X. H. Yu, J. B. Bates, G. E. Jellison, F. X. Hart: A stable thin-film lithium electrolyte: Lithium phosphorus oxynitride, *J. Electrochem. Soc.* 144(2) (1997) 524–532.

[48] A. Schwöbel, R. Hausbrand, W. Jaegermann: Interface reactions between LiPON and lithium studied by *in situ* X-ray photoemission, *Solid State Ionics* 273 (2015) 51–54.

[49] J. B. Bates, N. J. Dudney, B. Neudecker, A. Ueda, C. D. Evans: Thin-film lithium and lithium-ion batteries, *Solid State Ionics* 135(1–4) (2000) 33–45.

[50] K.-F. Chiu, C. C. Chen, K. M. Lin, C. C. Lo, H. C. Lin, W.-H. Ho, C. S. Jiang: Lithium phosphorus oxynitride solid-state thin-film electrolyte deposited and modified by bias sputtering and low temperature annealing, *J. Vac. Sci. Technol. A Vac. Surf. Film* 28 (2010) 568–572.

[51] P. D. Mani, S. Saraf, V. Singh, M. R. Robert, A. Vijayakumar, S. J. Duranceau, S. Seal, K. R. Coffey: Ionic conductivity of bias sputtered lithium phosphorus oxy-nitride thin films, *Solid State Ionics* 287 (2016) 48–59.

[52] Y. Hamon, P. Vinatier, E. I. Kamitsos, M. Dussauze, C. P. E. Varsamis, D. Zielniok, C. Roesser, B. Roling: Nitrogen flow rate as a new key parameter for the nitridation of electrolyte thin films. *Solid State Ionics* 179 (2008) 1223–1226.

[53] B. Fleutot, B. Pecquenard, H. Martinez, A. Levasseur: Thorough study of the local structure of LiPON thin films to better understand the influence of a solder-reflow type thermal treatment on their performances, *Solid State Ionics* 206 (2012) 72–77.

[54] Y. Hamon, A. Douard, F. Sabary, C. Marcel, P. Vinatier, B. Pecquenard, A. Levasseur: Influence of sputtering conditions on ionic conductivity of LiPON thin films, *Solid State Ionics* 177 (2006) 257–261.

[55] B. Wang, B. S. Kwak, B.C. Sales, J. B. Bates: Ionic conductivities and structure of lithium phosphorus oxynitride glasses, *J. Non-Cryst. Solids* 183 (1995) 297–306.

[56] L. Boukbir, R. Marchand, Y. Laurent, Z. J. Chao, C. Parent, G. Le Fleme: Investigation of the structure of phosphorus oxynitride glasses by local structural probes, *J. Less Common Met.* 148 (1989) 327–331.

[57] A. Le Sauze, L. Montagne, G. Palavit, F. Fayon, R. Marchand : X-ray photoelectron spectroscopy and nuclear magnetic resonance structural study of phosphorus oxynitride glasses, 'LiNaPON', *J. Non-Cryst. Solids* 263–264 (2000) 139–145.

[58] M. A. Carrillo, Solano, M. Dussauze, P. Vinatier, L. Croguennec, E. I. Kamitsos, R. Hausbrand, W. Jaegermann: Phosphate structure and lithium environments in lithium phosphorus oxynitride amorphous thin films, *Ionics* 22 (2016) 471–481.

[59] S. Jacke, J. Song, L. Dimesso, J. Brötz, D. Becker, W. Jaegermann: Temperature dependent phosphorous oxynitride growth for all-solid-state batteries, *J. Power Sources* 196 (2011) 6911–6914.

[60] F. Muñoz, A. Durán, L. Pascual, L. Montagne, B. Revel, A. C. Rodrigues: Increased electrical conductivity of LiPON glasses produced by ammonolysis, *Solid State Ionics* 179 (2008) 574–579.

[61] N. Mascaraque, J. L. G. Fierro, A. Duran, F. Muñoz: An interpretation for the increase of ionic conductivity by nitrogen incorporation in LiPON oxynitride glasses, *Solid State Ionics* 233 (2013) 73–79.

[62] Y. Su, J. Falgenhauer, A. Polity, T. Leichtweiss, A. Kronenberger, J. Obel, S. Zhou, D. Schlettwein, J. Janek, B. K. Meyer: LiPON thin films with high nitrogen content for application in lithium batteries and electrochromic devices prepared by RF magnetron sputtering, *Solid State Ionics* 282 (2015) 63–69.

[63] P. Birke, W. Weppner: Electrochemical analysis of thin film electrolytes and electrodes for application in rechargeable all solid state lithium microbatteries, *Electrochim. Acta* 42 (1997) 3375–3384.

[64] M. Dussauze, E. I. Kamitsos, P. Johansson, A. Matic, C. P. E. Varsamis, D. Cavagnat, P. Vinatier, Y. Hamon: Lithium ion conducting boron-oxynitride amorphous thin films: Synthesis and molecular structure by infrared spectroscopy and density functional theory modeling, *J. Phys. Chem. C* 117 (2013) 7202–7213.

[65] Y. Yoon, C. Park, J. Kim, D. Shin: The mixed former effect in lithium borophosphate oxynitride thin film electrolytes for all-solid-state micro-batteries, *Electrochimi. Acta* 111 (2013) 144–151.

[66] S.-J. Lee, J.-H. Bae, H.-W. Lee, H.-K. Baik, S.-M. Lee: Electrical conductivity in Li–Si–P–O–N oxynitride thin-films, *J. Power Sources* 123 (2003) 61–64.

[67] J. H. Kennedy, Z. Zhang: Improved stability for the SiS_2–P_2S_5–Li_2S–LiI glass system, *Solid State Ionics* 28–30 (1988) 726–728.

[68] S. Kondo, K. Takada, Y. Yamamura: New lithium ion conductors based on Li_2S–SiS_2 system, *Solid State Ionics* 53–56 (1992) 1183–1186.

[69] A. Hayashi, M. Tatsumisago, T. Minami: Electrochemical Properties for the Lithium Ion Conductive $(100-x)(0.6Li_2S \cdot 0.4SiS_2) \cdot xLi_4SiO_4$ Oxysulfide Glasses, *J. Electrochem. Soc.* 146 (1999) 3472–3475.

[70] T. Minami, A. Hayashi, M. Tatsumisago: Preparation and characterization of lithium ion-conducting oxysulfide glasses, *Solid State Ionics* 136–137 (2000) 1015–1023.

[71] A. Hayashi, H. Yamashita, M. Tatsumisago: Characterization of Li_2S–SiS_2–Li_xMO_y (M = Si, P, Ge) amorphous solid electrolytes prepared by melt-quenching and mechanical milling, *Solid State Ionics* 148 (2002) 381–389.

[72] P. Knauth: Inorganic solid Li-ion conductors: an overview, *Solid State Ionics* 180 (2009) 911–916.

[73] M. Tatsumisago, M. Nagao, A. Hayashi: Recent development of sulfide solid electrolytes and interfacial modification for all-solid-state rechargeable lithium batteries, *J. Asian Ceramics Societies* 1 (2013) 17–25.

[74] H. Muramatsu, A. Hayashi, T. Ohtomo, S. Hama, M. Tatsumisago: Structural change of Li_2S–P_2S_5 sulfide solid electrolytes in the atmosphere, *Solid State Ionics* 182 (2011) 116–119.

[75] M. Tachez, J.P. Malugani, R. Mercier, G. Robert: Ionic conductivity and phase transition in lithium thiophosphate Li_3PS_4, *Solid State Ionics* 14 (1984) 181–185.

[76] T. Ohtomo, A. Hayashi, M. Tatsumisago, K. Kawamoto: Characteristics of the Li_2O–Li_2S–P_2S_5 glasses synthesized by the two-step mechanical milling, *J. Non-Cryst. Solids* 364 (2013) 57–61.

[77] A. Hayashi, H. Muramatsu, T. Ohtomo, S. Hama, M. Tatsumisago: Improved chemical stability and cyclability in Li_2S–P_2S_5–P_2O_5–ZnO composite electrolytes for all-solid-state rechargeable lithium batteries, *J. Alloy Compd.* 591 (2014) 247–2250.

[78] J. Fu: Superionic conductivity of glass-ceramics un the system Li_2O–Al_2O_3–TiO_2–P_2O_5, *Solid State Ionics* 96 (1997) 195–200.

[79] C. Davis III, J. C. Nino: Microwave processing for improved ionic conductivity in Li_2O–Al_2O_3–TiO_2–P_2O_5 glass-ceramics, *J. Am. Ceram. Soc.* 98 (2015) 2422–2427.

[80] P. Goharian, A. R. Aghaei, E. E. Yekta, S. Banijamali: Ionic conductivity and microstructural evaluation of $Li_2O–TiO_2–P_2O_5–SiO_2$ glass-ceramics, *Ceram. Intl.* 41 (2015) 1757–1763.

[81] B. B. Owens, P. M. Skarstad: Ambient temperature solid state batteries, *Solid State Ionics* 53–56 (1992) 665–672.

[82] X. Xu, Z. Wen, X. Wu, X. Yang, Z. Gu: Lithium ion-conducting glass-ceramics of $Li_{1.5}Al_{0.5}Ge_{1.5}(PO_4)_3–xLi_2O$ ($x = 0.0–0.20$) with Good Electrical and electrochemical properties, *J. Am. Ceram. Soc.* 90 (2007) 2802–2806.

[83] H. S. Jadhav, M.-S. Cho, R. S. Kalubarme, J.-S. Lee, K.-N. Jung, K.-H. Shin, C.-J. Park: Influence of B_2O_3 addition on the ionic conductivity of $Li_{1.5}Al_{0.5}Ge_{1.5}(PO_4)_3$ glass ceramics, *J. Power Sources* 241 (2013) 5022–508.

[84] F. Mizuno, A. Hayashi, K. Tadanaga, M. Tatsumisago: New lithium-ion conducting crystal obtained by crystallization of the ion-conductive crystals precipitated from $Li_2S–P_2S_5$ glasses, *Adv. Mater.* 17 (2005) 918–921.

[85] S. Choi, M. Eom, C. Park, S. Son, G. Lee, D. Shin: Effect of Li_2SO_4 on the properties of $Li_2S–P_2S_5$ glass-ceramic solid electrolytes, *Ceram. Intl.* 42 (2016) 6738–6742.

[86] A. Hayashi, K. Minami, S. Ujiie, M. Tatsumisago: Preparation and ionic conductivity of $Li_7P_3S_{11-z}$ glass-ceramic electrolytes, *J. Non-Cryst. Solids* 356 (2010) 2670–2673.

[87] S. Ujiie, A. Hayashi, M. Tatsumisago: Preparation and ionic conductivity of $(100–x)(0.8Li_2S \cdot 0.2P_2S_5) \cdot xLiI$ glass-ceramics electrolytes, *J. Solid State Electrochem.* 17 (2013) 675–680.

[88] Y. Seino, T. Ota, K. Takada, A. Hayashi, M. Tatsumisago: A sulphide lithium super ion conductor is superior to liquid ion conductors for use in rechargeable batteries, *Energy Environ. Sci.* 7 (2014) 627–631.

[89] M. Kotobuki, M. Koishi: Sol–gel synthesis of $Li_{1.5}Al_{0.5}Ge_{1.5}(PO_4)_3$ solid electrolyte, *Ceram. Intl.* 41 (2015) 8562–8567.

[90] Y. Zhu, Y. Zhang, L. Lu: Influence of crystallization temperature on ionic conductivity of lithium aluminum germanium phosphate glass-ceramic, *J. Power Sources* 290 (2015) 123–129.

[91] S. Soman, D. Sonigra, A.R. Kulkarni: Isothermal crystallization and effect of soak time on phase evolution, microstructure and ionic conductivity of $Li_2O–Al_2O_3–TiO_2–P_2O_5$ glass–ceramic, *J. Non-Cryst. Solids* 439 (2016) 38–45.

[92] J. L. Narváez-Semanate, A. C. M. Rodrigues: Microstructure and ionic conductivity of $Li_{1+x}Al_xTi_{2-x}(PO_4)_3$ NASICON glass-ceramics, *Solid State Ionics* 181 (2010) 1197–1204.

[93] J. Fu: Fast Li^+ ion conduction in $Li_2O–Al_2O_3–TiO_2–SiO_2–P_2O_2$ glass-ceramics, *J. Am. Ceram. Soc.* 80 (1997) 1901–1903.

[94] H. Aono, E. Sugimoto, Y. Sadaoka, N. Imanaka, G. Adachi: The electrical properties of ceramic electrolytes for $LiM_xTi_{2-x}(PO_4)_3 + yLi_2O$, M=Ge, Sn, Hf, and Zr systems, *J. Electrochem. Soc.* 140 (1993) 1827–1833.

[95] J. S. Thockhom, N. Gupta, B. Kumar: Superionic conductivity in a lithium aluminum germanium phosphate glass-ceramic, *J. Electrochem. Soc.* 155 (2008) A915–A920.

[96] F. Mizuno, A. Hayashi, K. Tadanaga, M. Tatsumisago: High lithium ion conducting glass-ceramics in the system Li_2S–P_2S_5, *Solid State Ionics* 177 (2006) 2721–2725.

[97] F. Mizuno, A. Hayashi, K. Tadanaga, M. Tatsumisago: New lithium-ion conducting crystal obtained by crystallization of the Li_2S–P_2S_5 glasses, *Electrochem. Solid-State Lett.* 8(11) (2005) A603–A606.

Chapter 7

Application in All-Solid-State Battery

It is well known that Li batteries are an attractive energy source for portable electronic devices (e.g. laptop computers, cameras, toys, mobile phones, and so on) and transport applications. However, there are still many issues, such as safety issue. All-solid-state battery with a ceramic Li-ion conductor as a solid electrolyte has the following potential advantages [1, 2]:

(i) absence of electrolyte leakage,
(ii) high safety because of absence of flammable liquid electrolyte,
(iii) absence of problems relating to vaporization of liquid electrolytes,
(iv) absence of phase transition of the electrolyte at low temperature, which improves low temperature performance, and
(v) ease of miniaturization.

In addition, all-solid-state batteries usually have excellent storage stability and long life due to suppression of side reactions because only Li-ion can move inside the solid electrolyte.

In order to employ the ceramic Li-ion conductor in the high energy density Li secondary batteries as an electrolyte, the following electrical and physical properties should be fulfilled [3]:

(i) high lithium ion conductivity at operating temperature (preferably at room temperature), and negligible electronic conductivity,

(ii) negligibly low/or no grain-boundary resistance,

(iii) high stability against chemical and electrochemical reactions with both electrodes (cathode and anode), especially with elemental Li or Li-alloy-negative electrode during the preparation and operation of the cell,

(iv) compatible coefficient of thermal expansion with the electrodes,

(v) wide range of electrochemical decomposition potential (higher than 5.5 V vs. Li), and

(vi) environmental benignity, non-hygroscopic, low cost and ease of preparation.

The all-solid-state batteries can be categorized into two systems, i.e. thin-film type of battery and bulk type of battery (Fig. 7.1). The thin-film battery is composed of thin films of cathode, anode and solid electrolyte. The thickness of each layer is several 100 nm to several μm. Thus, thickness of whole battery is about 15 μm [4].

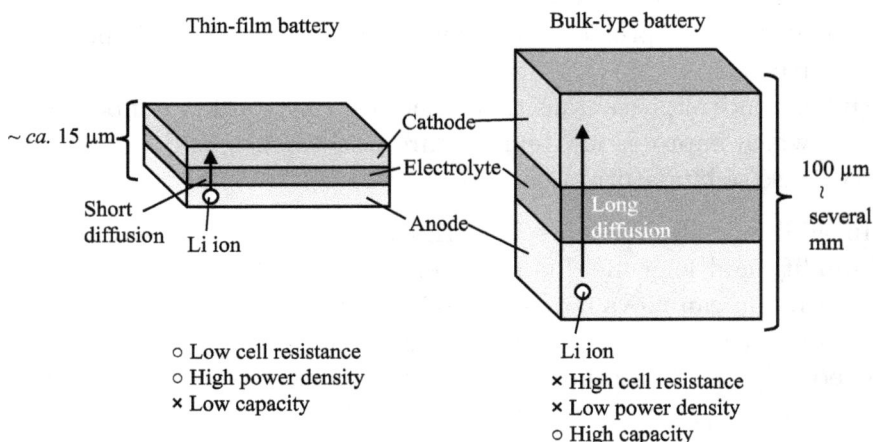

Fig. 7.1   Thin-film and bulk-type all-solid-state batteries.

This thinness of each battery component reduces cell resistance due to the short diffusion length of Li-ions, thus enabling the provision of large current flow. However, the capacity of the thin-film battery is low due to the loading of small amounts of active materials. The bulk type of batteries have thick electrodes ($>100\,\mu$m) which can provide a large capacity. However, the thick electrodes and electrolyte make the Li diffusion path longer, resulting in increase of the cell resistance. Therefore, it is usually difficult to obtain large current from the bulk type of batteries.

7.1   Thin-Film Battery

The thin-film battery is able to offer high power density, long lifetime, flexibility and is extremely lightweight, which tends to reduce the overall weight of the system [5]. Figure 7.2 shows a typical cross-section structure of the thin-film Li battery [5]. Physical vapor deposition (PVD) techniques, such as RF sputtering [6], bias sputtering [7], a combination of pulsed laser deposition (PLD) with pulse laser ablation [8] and so on have been used to prepare the thin-film battery by deposition of cathode, electrolyte and anode, in this order or vice versa. In these techniques, the deposition process proceeds at the atomic scale. Therefore, electrodes and electrolyte are in contact each other and so the resistance at electrode/electrolyte interface becomes lower, resulting in reduction of internal cell resistance [6]. The as-deposited thin films (electrodes and electrolyte) by the PVD are usually in an amorphous state. Therefore, amorphous electrode materials and electrolytes are preferred for the thin-film batteries because annealing of deposited amorphous thin films

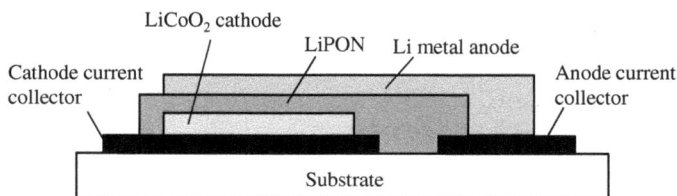

Fig. 7.2   Schematic cross-section of thin film battery [5].

at high temperature to obtain favorable structure and improve crystallinity would cause undesired reactions between electrodes, electrolyte and substrates. Normally, cathode materials require high crystallinity to offer high electrochemical performance. For example, $LiCoO_2$ cathode with hexagonal layered structure is a favorable structure for intercalation of Li-ions. However, $LiCoO_2$ cathode with a good crystallinity can be obtained after annealing the amorphous $LiCoO_2$ at above 700°C [9]. Accordingly, the cathode is usually deposited first on a substrate followed by annealing or is deposited on heated substrate, after which, amorphous electrolyte and anode are deposited. Furthermore, the Li-free thin-film battery, in which only cathode and electrolyte are deposited on a substrate and Li metal anode is plated during first charging, was suggested to survive solder flow conditions by Neudecker *et al.* [10], where $LiCoO_2$ cathode and LiPON electrolyte and Cu current collector were deposited by RF sputtering (Fig. 7.3). The Li metal anode was plated between LiPON and the Cu current collector in the first charging. By eliminating Li metal in as-prepared battery, the safety of the thin-film battery is largely improved. Until now, several amorphous oxide solid electrolytes have been used to prepare the thin-film batteries such as $Li_{3.6}Si_{0.6}P_{0.4}O_4$ [11–13], $LiBO_2$ [11], $Li_{6.1}V_{0.61}Si_{0.39}O_{5.36}$ [14] and so on. Among them, LiPON has been the most widely used electrolyte for the thin-film batteries. The LiPON-based thin-film all-solid-state battery was already commercialized by CYMBET corporation [15] and some of the others. Various kinds of electrode materials have been studied on thin-film batteries using LiPON electrolyte, such as V_2O_5 [16], $LiMn_2O_4$ [17–19], $LiCoO_2$ [5] and $Li_x(Mn_yNi_{1-y})_{2-x}O_2$ [20, 21] as cathode and Li metal [5, 17, 22], Si_3N_4 [23], SnO [14], Si [6] and amorphous Si–Sn oxynitride [24] as anode. Capacities of the thin-film battery with LiPON electrolyte were reported to be 45–60 μAh cm^{-2}-μm^{-1} [5, 7, 8]. Remarkable cycle stability of thin-film battery using LiPON was also reported. The cycling test over 30,000 cycles of $Li/LiPON/LiCoO_2$ system demonstrated that capacity fading was 0.0001% to 0.002% per cycle (Fig. 7.4) [5]. The thin-film form of crystalline oxide solid electrolytes has also been prepared for application of the thin-film batteries. For example, LLZ

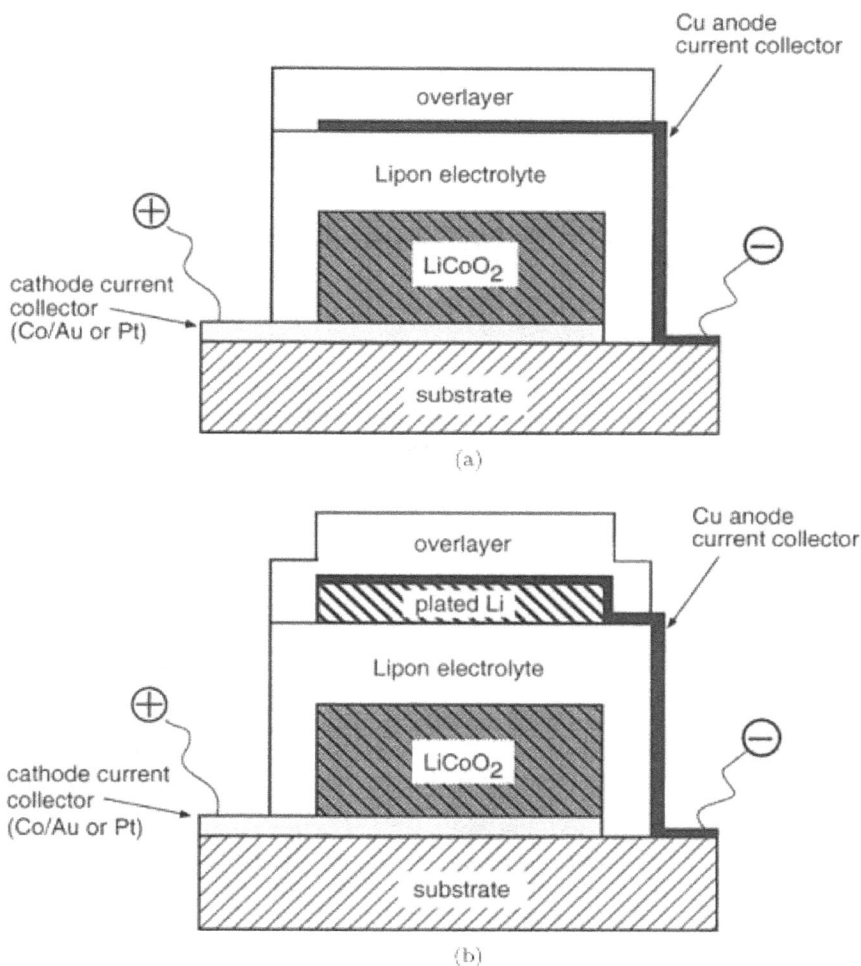

Fig. 7.3 Li-free thin film battery: (a) prior to initial charge and (b) after first charging.

prepared by pulsed laser deposition (PLD) [25], RF sputtering [26], aerosol deposition [27] and sol–gel method [28], and LATP(N) by RF sputtered in N_2 atmosphere [29] were reported. However, the ionic conductivity of these thin films is more than 2–3 orders lower than that in the bulk crystal phase (Table 7.1).

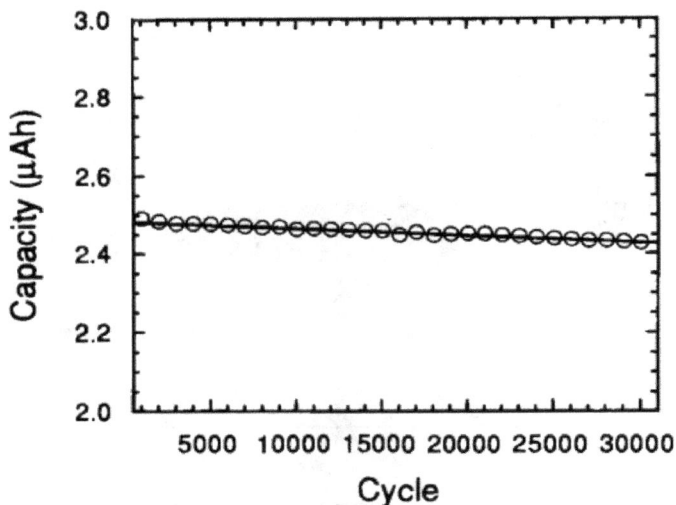

Fig. 7.4   Cycle performance of Li/LiPON/LiCoO$_2$ thin film battery [5].

Table 7.1   Room temperature conductivity of thin-film crystalline oxide solid electrolyte.

Electrolyte	Method	Conductivity (S cm^{-1})	Reference
LLZ	PLD	7.36×10^{-7}	[25]
LLZ	RF sputtering	4×10^{-7}	[26]
LLZ	Aerosol deposition	1.0×10^{-8} (140°C)	[27]
LLZ	Sol–gel	1.67×10^{-8}	[28]
LATP(N)	RF sputtering	1.2×10^{-5}	[29]

Unlike other types of oxide conductors showing generally 2 or 3 orders lower conductivity than their bulk counterparts, LLT thin films reported to date exhibit comparable or even higher conductivity than that in its bulk crystal form. The LLT thin films have been successfully prepared by various techniques, e.g. PLD [30–33], RF magnetron sputtering [34], e-beam evaporation [35], sol–gel [36, 37] and, more recently, by ALD [38]. A list of selected reports on LLT thin films is summarized in Table 7.2.

Table 7.2 Comparison of different deposition methods reported for LLT thin–film synthesis (a and c represent for amorphous and crystalline, respectively).

Synthesis techniques	Composition	Ionic conductivity (S/cm)	Activation Energy (eV)	Film thickness	Ref.
PLD	$a-Li_{0.5}La_{0.5}TiO_3$	1.13×10^{-5}	NA	400 nm	[33]
PLD	$a-Li_{0.5}La_{0.5}TiO_3$	2.0×10^{-5}	NA	360 nm	[32]
PLD	$a-Li_{0.45}La_{0.48}TiO_3$	8.75×10^{-4}	0.35	$0.56\,\mu m$	[31]
PLD	$a-Li_{0.5}La_{0.5}TiO_3$	3×10^{-4}	0.28	$1.2\,\mu m$	[39]
E-beam	$a-Li_{0.5}La_{0.4}TiO_3$	1.8×10^{-7}	0.32	800 nm	[35]
Sol-gel	$c-Li_{0.5}La_{0.5}TiO_3$	10^{-8}–10^{-7}	0.63–1.05	0.2–$1\,\mu m$	[37]
RF-sputtering	$c-Li_{0.24}La_{0.7}TiO_3$	—	—	300 nm	[34]
ALD	$c-Li_{0.32}La_{0.3}TiO_z$	—	—	<100 nm	[38]

The results show a notable difference in ionic conductivity for the same composition but different deposition conditions suggesting that the controlling of deposition parameters, e.g. temperature, time, background pressure and atmosphere, is critical to achieve high-quality LLT thin films as well as other types of electrolytes. Interestingly, these works also indicate that amorphous LLT thin films exhibit higher ionic conductivity than that of polycrystalline bulk LLT. This better conduction behavior might be attributed to the absence of grain boundaries, the open disordered structure, or stress in the film [31]. Currently, the highest ionic conductivity value reported to date is as high as 8.75×10^{-4} S/cm for amorphous $Li_{0.45}La_{0.48}TiO_3$ synthesized by PLD [31], which is two orders of magnitude higher than that of the widely used LiPON electrolyte in thin-film battery. Nonetheless, these LLT thin films still suffer from instability with lithium metal. Therefore, a barrier layer, such as LiPON and garnet electrolyte films acting as a protective layer, is essential to prevent the direct contact between LLT and metallic lithium anode.

Thin film of sulfide glass has been also studied. Li_2S–Ge_2S, Li_2S–P_2S_5 and Li_4GeS_4–Li_3PS_4 glasses were deposited by PLD, and their ionic conductivities range between 1.8×10^{-4} and 1.8×10^{-3} S cm^{-1} at room temperature [40–42]. However, these studies have not yet been applied to the thin-film batteries.

7.2 Bulk-Type All-Solid-State Li Battery

When compared with the thin-film battery, the bulk type of all-solid-state batteries can load a large amount of active materials and be scaled up easily. Therefore, it has been expected to serve as a power source for hybrid and electric vehicles.

Different from thin-film batteries, the bulk type of all-solid-state batteries are usually self-supported by thick electrodes and electrolyte. Thin-film type of electrolyte can also be used. In this case, electrodes must possess sufficient mechanical strength for self-supporting. The thick electrodes lengthen Li-ion and electron diffusion paths, causing an increase in the cell resistance (Fig. 7.5(a)).

Additionally, a poor solid–solid contact between the electrodes and the electrolyte can also lead to high cell resistance (Fig. 7.5(b)), which is unfavorable for obtaining high performance, especially under high current conditions, even though the solid electrolyte possesses high Li-ion conductivity. Figure 7.5(c) shows a SEM image of the $LiMn_2O_4$ electrode/LLT electrolyte interface. The interface

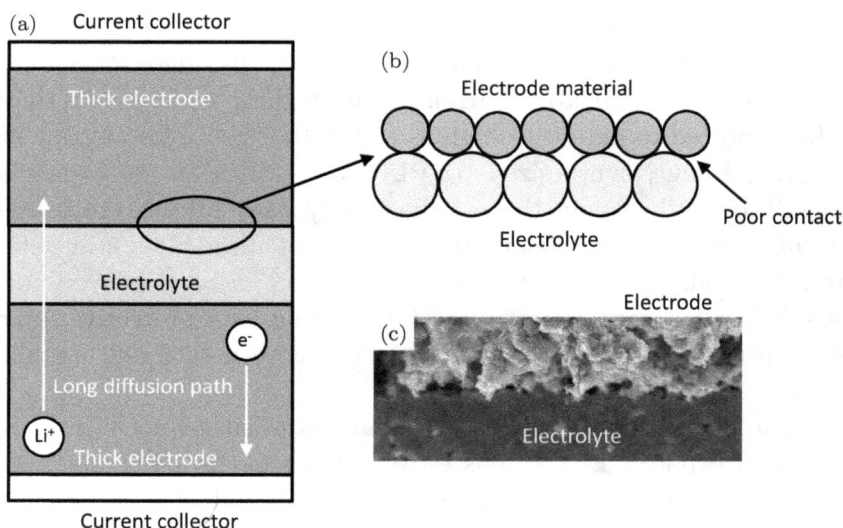

Fig. 7.5   Electrode–electrolyte interface of bulk-type all-solid-state battery.

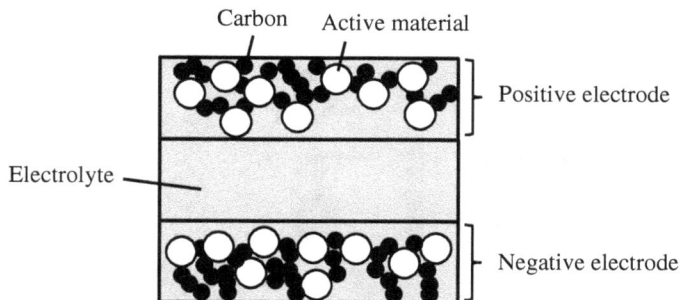

Fig. 7.6   Bulk-type all-solid-state battery with composite electrodes [43].

was prepared by dropping a precursor sol for electrode on the solid electrolyte, followed by calcination. Some gaps between the electrode and the electrolyte can be observed. As such, novel architectures are needed. The bulk-type all-solid-state battery with composite electrodes was suggested to reduce resistance in the thick electrode by Aboulaich *et al.* [43] (Fig. 7.6). In this configuration, the electrolyte layer is composed of active material, carbon and electrolyte to obtain a good contact between the active material and electrolyte, and the carbon also provides an electrical conduction path in the electrode layers.

The composite electrode contains electrode material, ionic conductive material and electrical conductive material. The electrode material should be loaded as much as possible to achieve high capacity, while the amount of ionic and electrical conductive materials should be minimized, but with enough amount possible to ensure ionic and electrical conduction. Based on this consideration, $LiFePO_4 / Li_{1.5}Al_{0.5}Ge_{1.5}(PO_4)_3 / Li_3V_2(PO_4)_3$ and $Li_3V_2(PO_4)_3/ Li_{1.5}Al_{0.5}Ge_{1.5}(PO_4)_3/Li_3V_2(PO_4)_3$ cells with several hundred micrometer thickness of composite electrode were prepared by spark plasma sintering (SPS). The cells exhibited a reversible capacity of $80\,mAh\,g^{-1}$ at a high temperature and a low C rate.

To reduce the interface resistance, 3D batteries, in which electrodes and electrolyte interfaces are enlarged three dimensionally, have also been suggested. In the 3D batteries, a large contact area between electrodes and electrolyte can be obtained, thus enabling

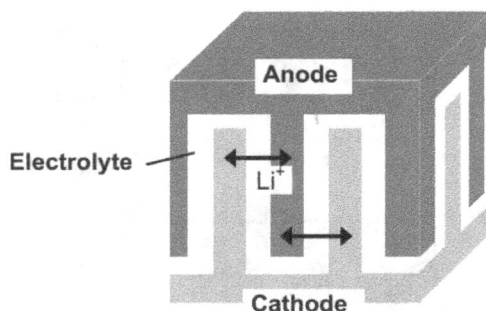

Fig. 7.7   Honeycomb-type 3D battery [44].

Fig. 7.8   SEM images of honeycomb LLT electrolyte (a) surface and (b) cross-section.

a reduction of the resistance at the interface between electrodes and electrolyte. Honeycomb-type battery was suggested as one of the 3D battery configurations (Fig. 7.7) [44]. This configuration can not only provide large contact area but also reduce Li-ion transportation length in the electrolyte. The honeycomb-type solid electrolyte with hole size of $180 \times 180 \times 180\,\mu m$ was prepared (Fig. 7.8). $LiCoO_2$ cathode and $Li_4Mn_5O_{12}$ anode were impregnated with their precursor sol on each side of the holes followed by calcination. When a dispersion of $LiCoO_2$ in ethanol was used for the impregnation, a gap was formed between the cathode and the electrolyte layer (Fig. 7.9(a)). This implies that the bottom side of cathode layer does not contribute to the Li-ion transportation.

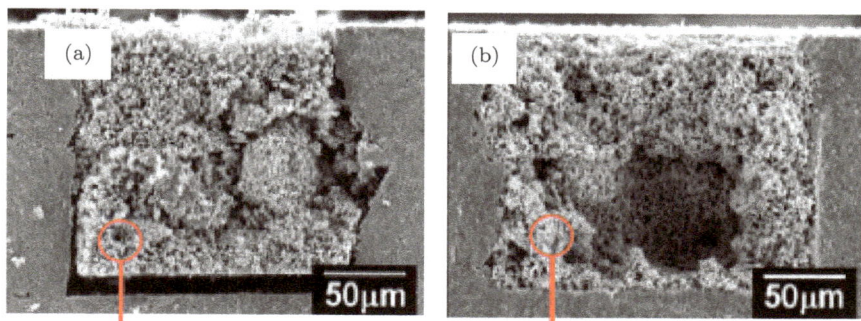

Fig. 7.9   Cross-sectional SEM images of the half-cells prepared by impregnation of (a) ethanol suspension of $LiCoO_2$ particles and (b) $LiCoO_2$ particles/precursor sol mixture.

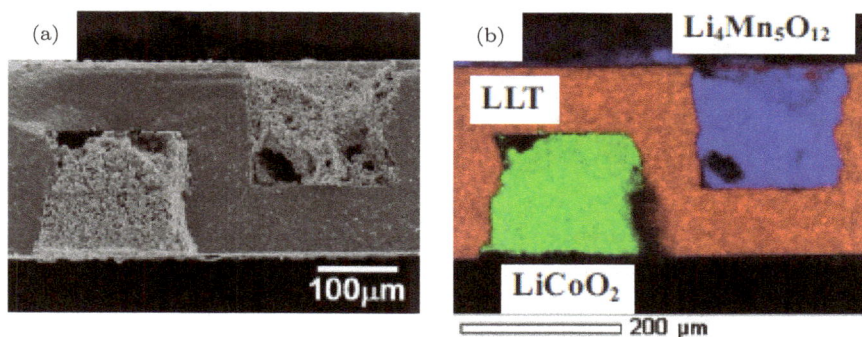

Fig. 7.10   Cross-sectional SEM image (a) and EDS mapping (b) of $LiCoO_2$/honeycomb LLT/$Li_4Mn_5O_{12}$ cell.

It was noted that the gap disappeared when using the precursor sol for the disperse media (Fig. 7.9(b)). If $LiCoO_2$ cathode and $Li_4Mn_5O_{12}$ anode are used, $LiCoO_2$ should be impregnated and calcined first because calcination temperature for $LiCoO_2$ (800°C) is higher than that of $Li_4Mn_5O_{12}$ (700°C). By preparing the $LiCoO_2$ cathode first, undesired reactions that may proceed at the anode side during calcination at high temperature can be possibly avoided. The $LiCoO_2$/honeycomb LLT/$Li_4Mn_5O_{12}$ cell was prepared successfully by appropriate impregnation and calcination procedures (Fig. 7.10). The honeycomb-type battery with LLT electrolyte, $LiCoO_2$ cathode

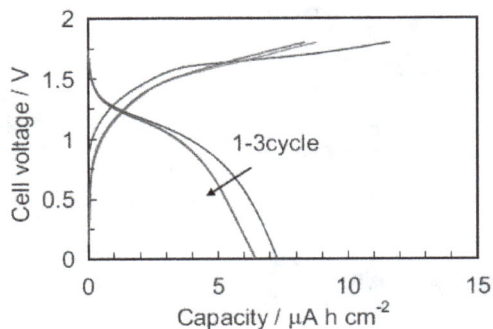

Fig. 7.11   Charge and discharge curves of $LiCoO_2$/honeycomb LLT/$Li_4Mn_5O_{12}$ cell.

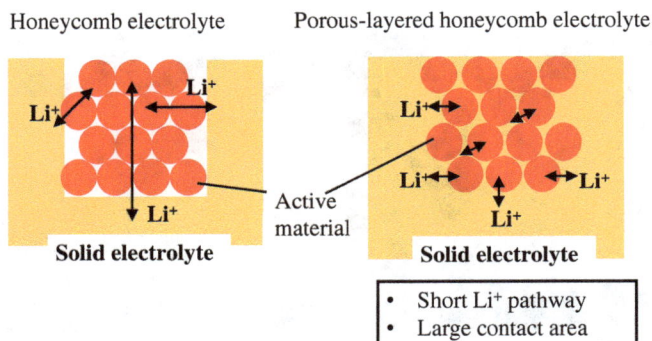

Fig. 7.12   Honeycomb battery with porous structure in its holes.

and $Li_4Mn_5O_{12}$ anode demonstrated charge and discharge behavior at room temperature (Fig. 7.11). Furthermore, the performance of the honeycomb battery could be improved by building a porous structure in the honeycomb holes (Fig. 7.12) [45]. Fabrication of porous LLT in the hole of the honeycomb LLT is another way to increase the already enlarged contact area between $LiMn_2O_4$ cathode and solid electrolyte and reduce the interfacial resistance. Figure 7.13 shows a comparison of the impedance of porous honeycomb LLT with that of non-porous honeycomb LLT. The evidence of reduction of impedance can be seen in Fig. 7.13 where a semicircle at 120 Hz, which can be assigned to interfacial resistance between honeycomb

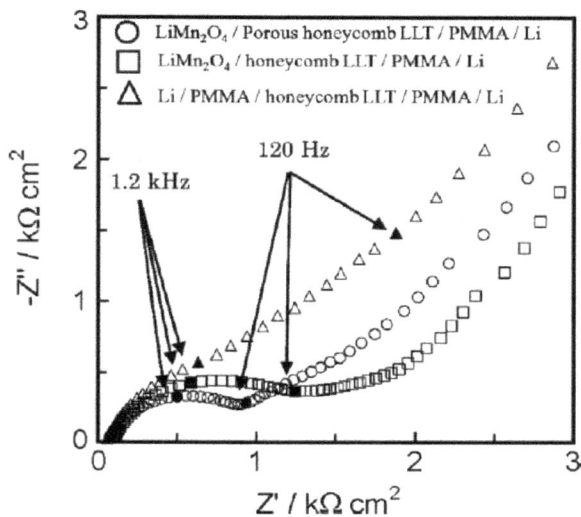

Fig. 7.13 Complex impedance plot of $LiMn_2O_4$/honeycomb LLT and $LiMn_2O_4$/porous honeycomb LLT.

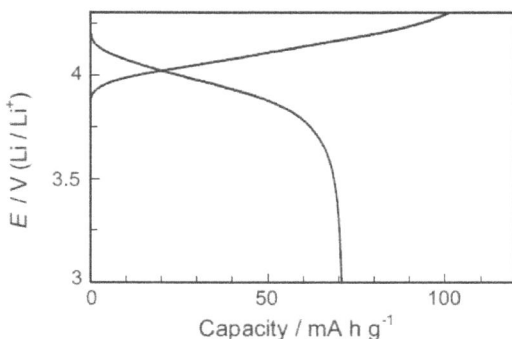

Fig. 7.14 Charge and discharge curves of $LiMn_2O_4$/porous honeycomb LLT cell.

electrolyte and $LiMn_2O_4$, becomes smaller on using the porous honeycomb electrolyte. The discharge capacity of the porous honeycomb battery is $71\,mA\,h\,g^{-1}$ (Fig. 7.14), whereas the discharge capacity of the $LiMn_2O_4$/honeycomb cell is only $1.27\,mA\,h\,g^{-1}$. The porous structure significantly contributes to the high discharge capacity, confirming that enlargement of contact area between electrode and

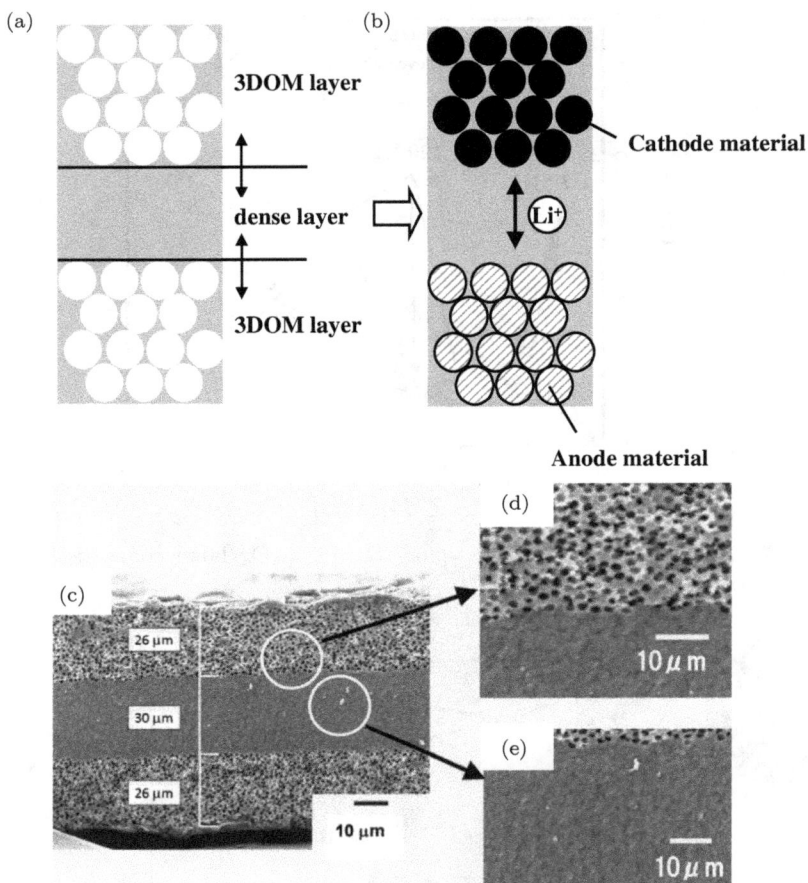

Fig. 7.15   Structure and cross-sectional SEM images of 3DOM battery [46].

electrolyte is critical to obtain high performance all-solid-state Li batteries.

As another example, 3D ordered microporous (3DOM) structure was also applied for the 3D battery (Fig. 7.15) [46]. The 3DOM battery was prepared by the colloidal crystal templating method. Two 3DOM porous layers about $30\,\mu$m in thickness were separated by a dense layer (about $30\,\mu$m thickness). Then, each porous layer was filled with cathode and anode materials, respectively. Charge and

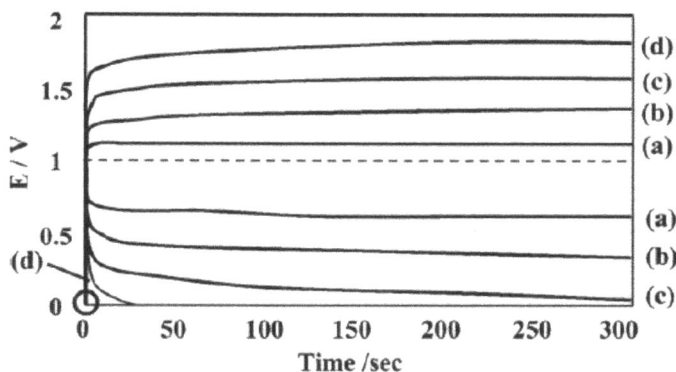

Fig. 7.16 Chronopotentiograms of $LiMn_2O_4$/3DOM LLT/$LiMn_2O_4$ cell: (a) 100 nA cm^{-2}, (b) 200 nA cm^{-2}, (c) 300 nA cm^{-2}, (d) 400 nA cm^{-2}.

discharge currents were observed on operation at room temperature, indicating that the 3DOM battery worked as a rechargeable battery (Fig. 7.16). As expected, the compatibility of electrode materials with electrolyte is important to build a high performance all-solid-state battery. In the bulk type of battery with oxide solid electrolyte, heat treatment (sintering) is used to obtain good contact between the electrode and the electrolyte. Therefore, electrode materials that do not react with the electrolyte under the sintering condition should be chosen.

The compatibility of $LiCoO_2$ and $LiMn_2O_4$ with LLT was studied [47]. Figure 7.17 shows XRD patterns of LLT/$LiCoO_2$ and LLT/$LiMn_2O_4$ after calcination. In $LiCoO_2$, impurities such as Co_3O_4 and $La_2Ti_2O_7$ are formed after sintering with LLT electrolyte. On the other hand, impurity formation was not observed in $LiMn_2O_4$. The impurity would act as resistance layer, resulting in increase of the resistance at the interface as evidenced by the complex impedance plots (Fig. 7.18). The $LiMn_2O_4$/LLT cell without impurity demonstrated better performance (Fig. 7.19). In the case of NASION-type oxide electrolytes, phosphate electrode materials [48] or those with NASICON structure [43, 49, 50] are more suitable to create the interface with low resistance. A similar NASICON structure between electrode materials and electrolyte would facilitate Li-ion transport

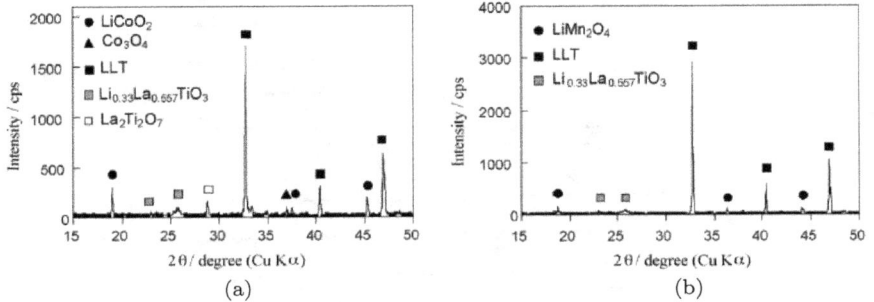

Fig. 7.17　XRD patterns of two types of batteries after calcination: (a) LiCoO$_2$/honeycomb LLT and (b) LiMn$_2$O$_4$/honeycomb LLT.

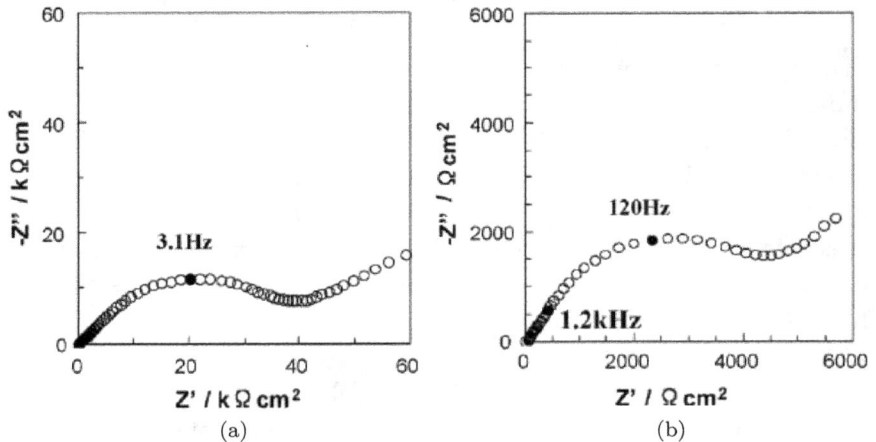

Fig. 7.18　Complex impedance plot of (a) LiCoO$_2$/honeycomb LLT and (b) LiMn$_2$O$_4$/honeycomb cells.

and reduce resistance at the interface. However, other electrodes such as LiCoO$_2$, LiMn$_2$O$_4$, and Li$_4$Ti$_5$O$_{12}$ are chemically unstable with NASICON type of electrolytes when compared with phosphate and NASICON type of electrode materials [51–53]. The procedure to prepare the electrodes is also key to reducing the resistance at the interface. LiMn$_2$O$_4$ electrode was prepared on the LLT electrolyte by the sol-gel method using different precursor sol compositions (Fig. 7.20) [54]. A gap was observed at the interface between

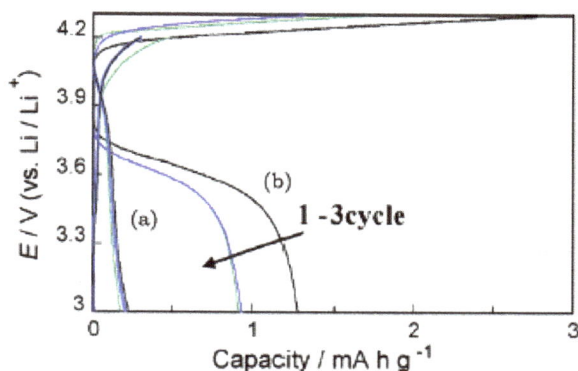

Fig. 7.19 Charge and discharge curves of (a) LiCoO$_2$/honeycomb LLT and (b) LiMn$_2$O$_4$/honeycomb cells.

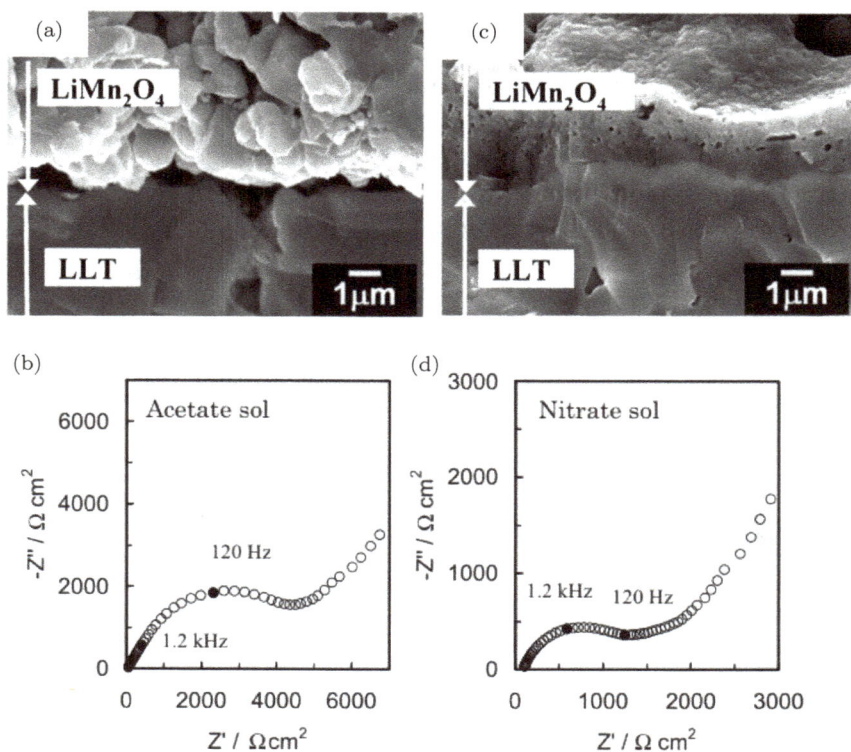

Fig. 7.20 SEM images of LiMn$_2$O$_4$/LLT interface prepared by (a) acetate sol and (c) nitrate sol. Complex impedance plots of LiMn$_2$O$_4$/LLT cell prepared by (b) acetate sol and (d) nitrate sol [56].

$LiMn_2O_4$ and LLT when acetate gel was used. On the other hand, the $LiMn_2O_4$ electrode prepared by using nitrate sol was firmly in contact with the LLT electrolyte. The resistance at the interface can also be reduced by using the nitrate sol.

Different from oxide solid electrolytes, for sulfide solid electrolytes, small grain boundary resistance and physically good contact can be obtained easily by simple cold press [55]. Hence, sintering process is not necessary. This feature is convenient for constructing bulk type of batteries, because the batteries can be simply assembled by pressing powders of battery components into a three-layered structure of anode/electrolyte/cathode. The high polarization of sulfide ions weakens the interaction between Li-ions and sulfide ions and facilitates fast Li-ion transport. However, this causes a problem when the sulfide electrolytes come into contact high-voltage cathodes like $LiCoO_2$. At the interface between the high-voltage cathode and sulfide electrolyte, Li-ion concentration on the solid electrolyte side is lowered to reach equilibrium because of the difference of chemical potential between the electrolyte and the electrode and the weak attraction of sulfide ions for Li-ions. As a consequence, a Li-depleted layer forms on the electrolyte side of the interface (Fig. 7.21). This layer is highly resistive because of the depletion of charge carriers. In this case, Li-ion conduction in the electrolyte is no longer a rate-determining step, but it is so in the Li-depleted layer. Therefore, a buffer layer to shield the sulfide electrolytes from the cathode is

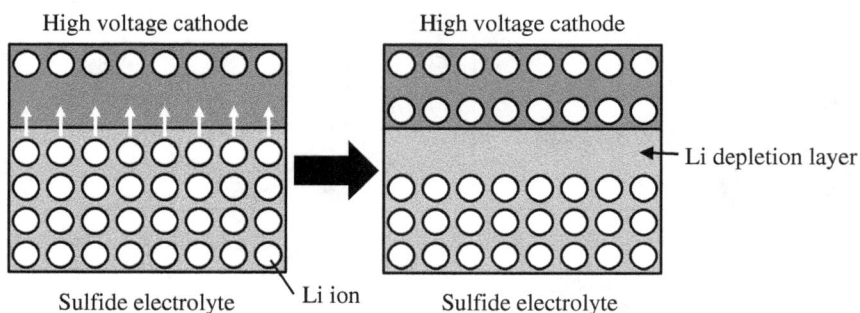

Fig. 7.21   Schematic image of Li depletion layer formation at the interface between high-voltage cathodes and sulfide electrolytes.

needed. The buffer layer should have high ionic conductivity. For this reason, oxide solid electrolyte can be used as the buffer layer.

When $LiNi_{0.5}Mn_{1.5}O_4$ particles are coated with Li_3PO_4 oxide electrolyte buffer layer, discharge capacity of the $In/80Li_2S$–$20P_2S_5/LiNi_{0.5}Mn_{1.5}O_4$ cell was dramatically improved [57]. Other oxides such as $Li_4Ti_5O_{12}$ [58–61], $LiNbO_3$ [62], $LiTaO_3$ [63] and so on have also been used with obvious improvement in battery performance. In Al-substituted $LiCoO_2$, Al-rich layer is naturally formed on the particle surface by annealing at 700°C. Figure 7.22

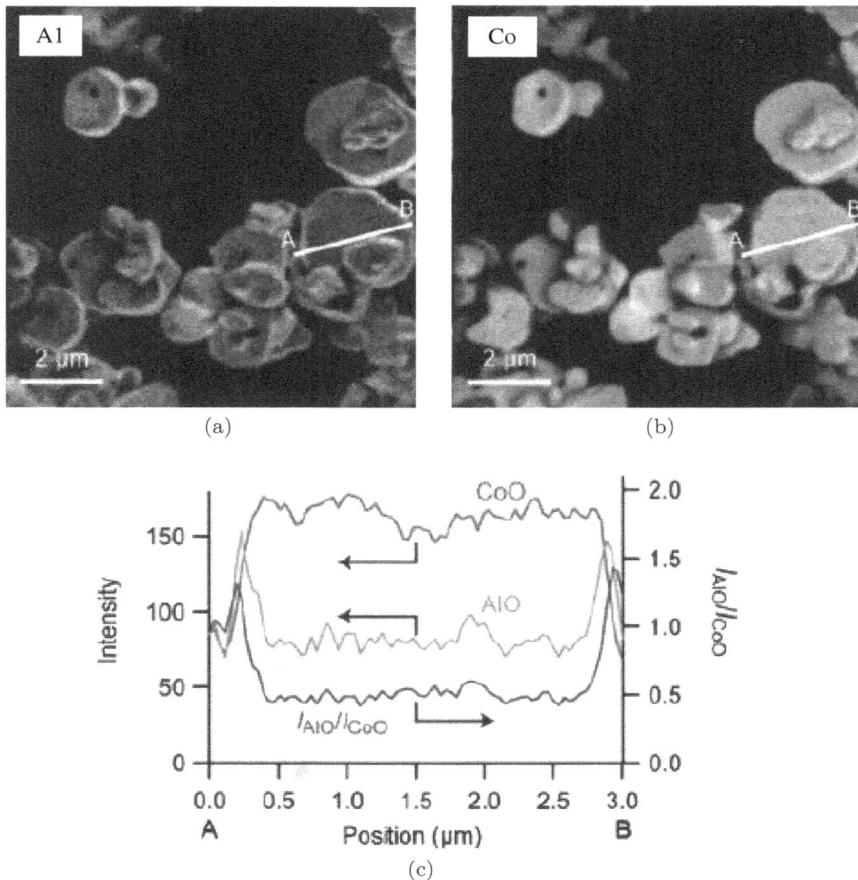

Fig. 7.22 Elemental mapping of $LiAl_{0.08}Co_{0.92}O_2$: (a) Al, (b) Co, (c) line analysis along A–B.

shows an elemental mapping of $LiCo_{0.92}Al_{0.08}O_2$ [64]. It can be clearly seen that concentration of Al is higher at the surface of the particle interface, while Co is lower. This finding implies that the Al-rich layer is naturally formed on the surface of $LiCo_{0.92}Al_{0.08}O_2$ particle by annealing. The formed Al-rich layer acts as a buffer layer. Therefore, the Al-substituted $LiCoO_2$ exhibits high rate capability even without the buffer layer (Fig. 7.23). Additionally, the buffer layer can suppress interdiffusion of elements between cathode and solid electrolyte that increases the electrode resistance, leading to long cycle life. A comparison of EDS mapping of Co element of the Li_2SiO_3-coated $LiCoO_2$ after initial charging with non-coated $LiCoO_2$ is shown in Fig. 7.24 [65]. The interface between Co and

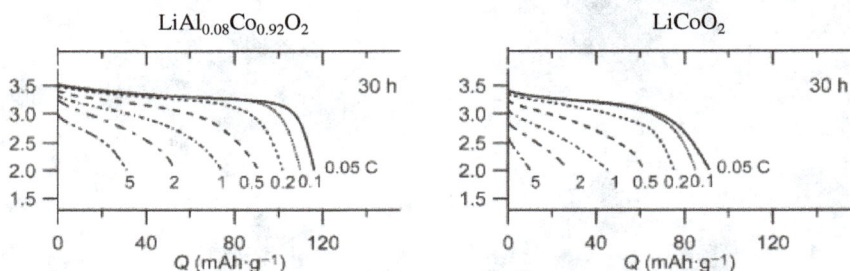

Fig. 7.23 Discharge curve of (a) $LiAl_{0.08}Co_{0.92}O_2$ and (b) $LiCoO_2$.

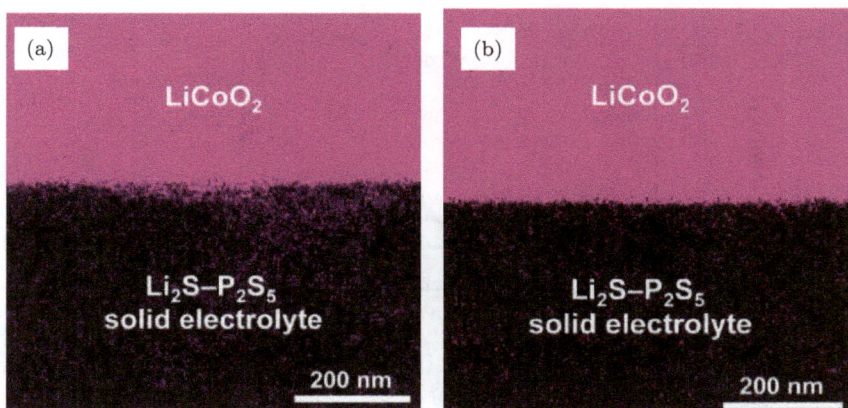

Fig. 7.24 Cross-sectional EDS mapping of Co element of (a) $LiCoO_2$ and (b) Li_2SiO_3-coated $LiCoO_2$ electrode after first charging.

the solid electrolyte layer was indistinct in the non-coated $LiCoO_2$, indicating that the diffusion of Co element occurred. On contrary, in the Li_2SiO_3-coated $LiCoO_2$, the interface was observed to be clear. From this example, it is clear that not only searching for a better buffer layer material but also formation of uniform buffer layer and control of its thickness are very important to develop the all-solid-state battery with sulfide solid electrolytes.

7.3 Li–S Battery

The typical lithium-sulfur battery consists of sulfur as the cathode, lithium as the anode, and liquid electrolyte with separator (Fig. 7.25). The sulfur cathode has a high theoretical capacity of $1672 \, mAh \, g^{-1}$, which is one order of magnitude higher than that of the conventional commercial cathode $LiCoO_2$, making the lithium–sulfur battery have an almost five times higher energy density compared with the conventional lithium-ion batteries (2500 vs. $500 \, Wh \, kg^{-1}$) [66]. Moreover, sulfur is inexpensive, non-toxic and abundant. The significant advances of lithium–sulfur battery are distinct, but commercial application is so far limited owing to some key challenges that must be addressed [67].

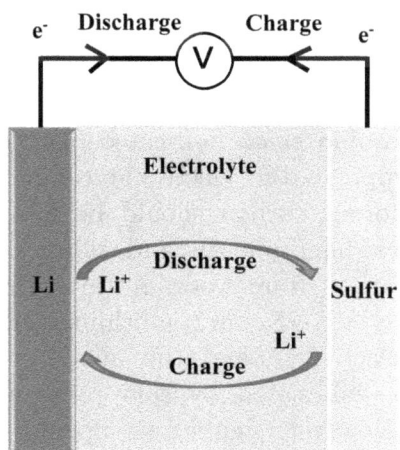

Fig. 7.25   Schematic illustration of a typical lithium-sulfur battery.

The first main issue is that sulfur is both ionically and electrically insulating (e.g. the electronic conductivity of sulfur is $5 \times 10^{-30} \, S \, cm^{-1}$ at room temperature) [68]. To overcome the insulating nature of sulfur, addition of conductive materials such as carbon, metals, and/or conducting polymers to make a composite with sulfur and reduction of sulfur particle size to shorten the diffusion path of electrons and lithium ions have been attempted [69–71]. The second main issue is that sulfur is reduced to lithium polysulfide intermediates that are highly soluble in organic liquid electrolytes [68]. These polysulfides, produced at the cathode side, diffuse through the electrolyte to the lithium anode side and were reduced to Li_2S and polysulfides with short chain. Subsequently, the reduced polysulfides diffuse back to the cathode side. This phenomenon is well described as the "polysulfide shuttle", whereby sulfur active mass is lost and capacity is decreasing. The migrated sulfur species lead to the redistribution of sulfur and generate the third main issue: poor cycle life [72]. The migrated sulfur species do not redeposit in their original positions upon cycling. Consequently, the microstructure of the sulfur cathode progressively changes with cycling, which leads to fast aging of electrodes and a quick capacity fading.

Thus, all-solid-state lithium–sulfur batteries with replacement of liquid electrolytes by ceramic electrolytes are developed to overcome the problem of liquid electrolyte lithium–sulfur battery. Furthermore, the all-solid-state lithium–sulfur batteries have advantages of high reliability and safety, which is important for large-scale application. Kanno *et al.* fabricated 3D mesoporous cathode configuration to improve the electronic conduction [73]. In this process, 3D mesoporous carbon should be first synthesized from a template of silica nanospheres, and sulfur, as a gas, is then diffused into the carbon framework. A thio-LISICON-type sulfide electrolyte $Li_{3.25}Ge_{0.25}P_{0.75}S_4$ and a lithium–aluminum composite are used as the electrolyte and the anode, respectively. Then, the battery can be fabricated by cold press with a pressure of around 500 MPa. However, significant capacity decrease was still observed. The low ionic conductivity of the composite electrode is one reason for the poor cyclability. To overcome the issues of poor reactivity and low ionic conductivity of sulfur, Chikusa *et al.* [74] prepared a composite cathode of sulfur, P_2S_5, carbon and solid electrolyte $Li_{1.5}PS_{3.3}$-LiI. The cathode powder and solid electrolyte $Li_{10}GeP_2S_{12}$ were pressed to pellet under 200 MPa. The Li–In alloy was attached on the side of $Li_{10}GeP_2S_{12}$ as anode. The solid electrolyte $Li_{10}GeP_2S_{12}$ was reported to possess ionic conductivity as high as liquid electrolyte [75]. The discharge capacity is around $1660\,mAh\,g^{-1}$ at $0.64\,mA\,cm^{-2}$. Furthermore, the all-solid-state lithium–sulfur battery with a composite cathode of sulfur–P_2S_5–carbon–$0.82(Li_{1.5}PS_{3.3})$–$0.18LiI$ shows excellent cycling performance over 100 cycles at $1.3\,mA\,cm^{-2}$. Furthermore, a solid-state lithium–sulfur battery was prepared by using Li_3PS_4 as solid electrolyte [76]. The cathode composite consisted of sulfur, carbon nanofibers and Li_3PS_4 with a weight ratio of 30:10:60. The battery achieved a capacity of $\sim 1600\,mAh\,g^{-1}$ at 0.05C (Fig. 7.26).

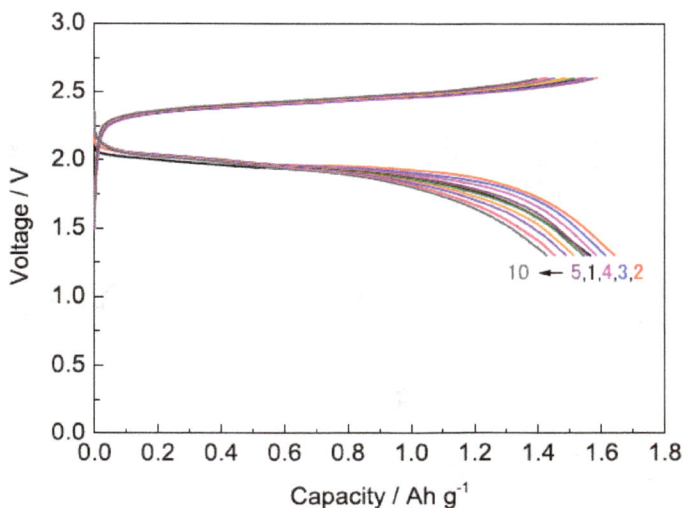

Fig. 7.26   Charge and discharge curves of $Li/Li_3PS_4/S$ cell [76].

The poor performance of all-solid-state lithium–sulfur battery could be mainly due to the poor ionic and electronic conductivity of sulfur. Therefore, both ionic and electronic conductivity of the sulfur cathode should be improved. Although composite cathodes generally are prepared from sulfur, conductive additives and solid electrolytes by ball milling method, the uniform distribution of these three components cannot be guaranteed. The bottom-up method has been used to synthesize the nanocomposite Li_2S–Li_6PS_5Cl–C [77]. At first, Li_2S active material, Li_6PS_5Cl solid electrolyte and PVP were dissolving together in ethanol, followed by drying at 100°C to evaporate the ethanol and to obtain a nanocomposite of Li_2S, Li_6PS_5Cl and PVP. The obtained nanocomposite was carbonized at 550°C in Ar. Figure 7.27 reveals SEM-EDS and TEM images of Li_2S–Li_6PS_5Cl–C composite. These three components are well mixed and uniformly dispersed, as can be observed in the SEM-EDS image. In the TEM image, a good contact of these three components is observed. The electronic and ionic conductivities of Li_2S–Li_6PS_5Cl–C are $2.2 \times 10^{-5}\,S\,cm^{-1}$ and $9.6 \times 10^{-6}\,S\,cm^{-1}$, respectively. The balanced electronic and ionic conductivities of the composite electrode would lead to better performance. The Li_2S–C electrode was only able to deliver a low initial capacity of

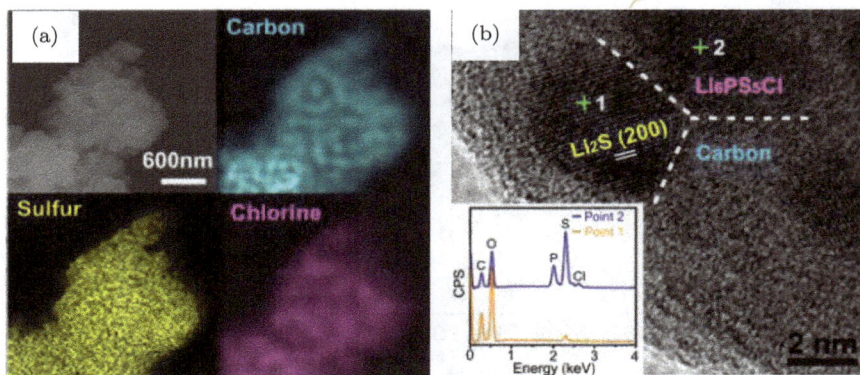

Fig. 7.27   (a) SEM-EDS and (b) TEM images of Li_2S–Li_6PS_5Cl–C [77].

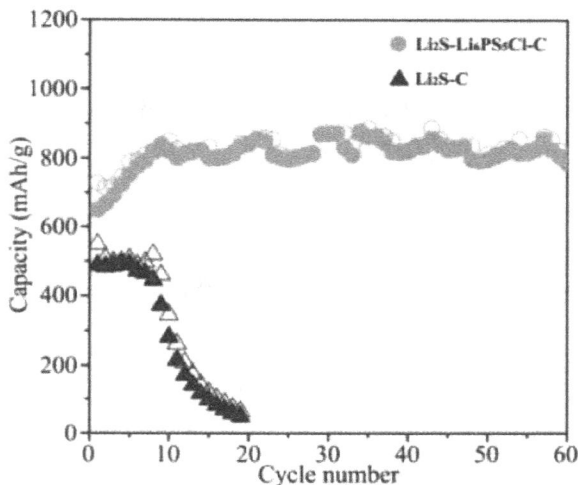

Fig. 7.28   Cycle performance of Li_2S–Li_6PS_5Cl–C and Li_2S–Li_6PS_5Cl [77].

489 mAh g^{-1} with quick decay to 49 mAh g^{-1} after just 10 cycles. On the contrary, the Li_2S–Li_6PS_5Cl–C electrode provided a high initial capacity of 648 mAh g^{-1}, and stabilizes at 830 mAh g^{-1} for 60 cycles (Fig. 7.28).

The replacement of liquid electrolyte by a solid one for the lithium–sulfur battery can eliminate the polysulfide shuttle and enables high energy density and improved safety in design. Some promising solid sulfide electrolytes with high ionic conductivities have been developed for all-solid-state lithium–sulfur battery. New composite electrodes with good ionic and electronic conductivities have also been developed. Though promising, there are still challenges for the all-solid-state lithium–sulfur battery. New solid electrolytes with sufficient ionic conductivities as high as liquid electrolytes should be developed, chemical compatibility of solid electrolytes with metallic lithium should be improved, new composite electrodes with both high ionic and electronic conductivities should be invented, interfacial resistances between solid electrolyte and electrodes should be decreased and, finally, large loading of active materials should be achieved.

7.4   Li–Air Battery

Metal–air batteries have received much attention among the researchers in the past decades. In the metal–air battery, anode is composed of pure metal which does not include any matrix components. Additionally, the metal–air battery does not need to load cathode material(s) because the interface between the electrolyte and air can be the cathode. These electrode configurations can reduce the weight of battery significantly, and thus the metal–air batteries possess extremely high energy density. The energy density of some metal–air batteries in shown Table 7.3 [78].

Among the many types of metal–air batteries, Li–air battery has been the most attractive one because of its high energy density, which is several fold higher than those of the present battery systems [79].

A typical Li–air cell is composed of Li metal as an anode, porous carbon as a cathode and an electrolyte in which Li-ions can move from one electrode to the other (Fig. 7.29(a)). Porous carbon cathode is employed to provide electronic conduction path and the porous structure helps air to easily access the electrolyte/porous carbon interface where the cathode reaction occurs. Depending on

Table 7.3   Theoretical energy density and operation voltage of various metal–air batteries.

Metal–air battery	Theoretical energy density ($Wh\,kg^{-1}$, excluding oxygen)	Theoretical energy density ($Wh\,kg^{-1}$, including oxygen)	Open circuit voltage (V)
Li–air	11,140	5,210	2.91
Na–air	2,260	1,677	2.3
Mg–air	6,462	2,789	2.93
Al–air	8,140	4,300	1.2
Zn–air	1,350	1,090	1.65
Ge–air	7,850	1,480	1
Fe–air	2,044	1,431	1.3
Ca–air	4,180	2,990	3.12
Si–air	9,036	4,217	1.6
K–air	1,700	935	2.48

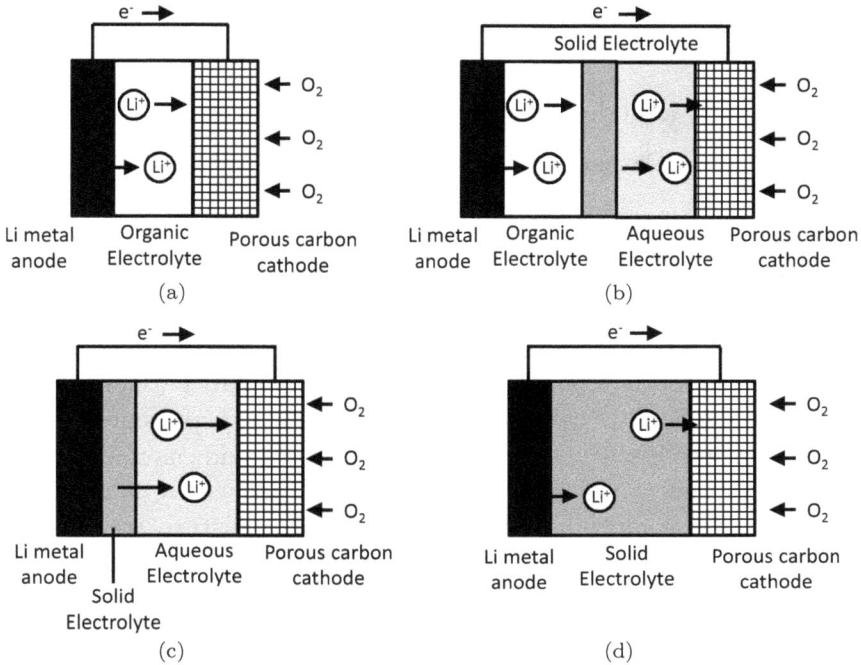

Fig. 7.29 Various kinds of batteries.

the property of the electrolyte, different electrochemical reactions proceed in the Li-air cell.

7.4.1 *Solid electrolytes for Li-air battery*

Li-air batteries can be categorized into four types depending on the property of the electrolyte (Fig. 7.29) [80]. The Li-ion conductive membrane (solid electrolyte) is included in three of four categories. As shown in Fig. 7.29(a), the Li-air battery consists of only liquid electrolyte, and does not contain solid electrolyte. This type of battery was first reported to be a rechargeable one by Abraham *et al.* in 1996 [81]. The Li metal anode is oxidized during the discharge process and releases Li-ions which diffuse to the cathode in the electrolyte. On the cathode, O_2 is reduced to O^{2-} or O_2^{2-} species. At the same time, the Li-ions react with these oxygen species and then form Li_2O and Li_2O_2 on the carbon cathode. The cell reaction is described as follows [82]:

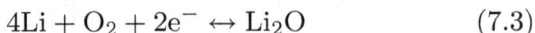

$$\text{Anode} \quad 2Li \leftrightarrow 2Li^+ + 2e^- \tag{7.1}$$

$$\text{Cathode} \quad 2Li^+ + O_2 + 2e^- \leftrightarrow Li_2O_2 \tag{7.2}$$

$$4Li + O_2 + 2e^- \leftrightarrow Li_2O \tag{7.3}$$

Reaction (7.2) is reversible, but reaction (7.3) is irreversible. Based on the reaction (7.2), the energy density including oxygen is estimated to be $3460 \, Wh \, kg^{-1}$ with an open circuit voltage of 2.96 V. To reduce overpotential of reactions (7.2) and (7.3), cobalt phthalocyanine had been used as a catalyst. Later, Bruce *et al.* presented high-performance cells with electrolytic manganese dioxide as the catalyst [83, 84]. However, the active life of these cells was still not long enough because oxygen, carbon dioxide and moisture go through the electrolyte and react with the Li anode. Although some positive results have been obtained in pure oxygen, dry oxygen and a mixture of oxygen and nitrogen [85, 86], air containing moisture should be used for practical applications. The battery performance is drastically reduced in air even though hydrophobic membranes are set to prevent incorporation of moisture from the cathode side. Thus, protection of Li metal anode becomes important to lengthen battery life [87].

The Li-air cell shown in Fig. 7.29(b) is composed of organic and aqueous electrolytes on Li metal and air sides, respectively, and a solid electrolyte is placed between them to avoid mixing both liquid electrolytes and to protect organic electrolyte and Li metal from water. The solid electrolyte should be stable in contact with both organic and aqueous electrolytes under battery operation and does not have pores to prevent infiltration of both liquid electrolytes. Figure 7.29(c) is similar to Fig. 7.29(b), but the organic electrolyte is eliminated and the solid electrolyte is in contact with the Li metal directly. In this case, the cell structure becomes simpler, although an additional property is required in the case of a solid electrolyte, i.e. high stability in contact with Li metal. Moreover, a solid–solid interface between Li metal and solid electrolyte would

Small contact area

Large contact area

Li metal   Solid electrolyte   Li metal   Liquid electrolyte

(a)   (b)

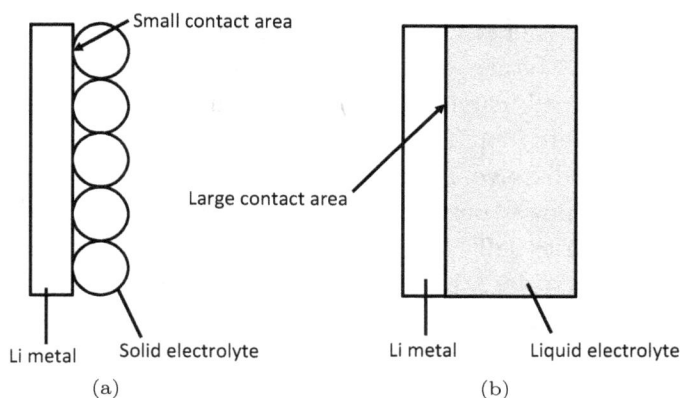

Fig. 7.30   Contact difference between liquid and solid electrolytes.

reduce the contact area, resulting in inferior battery performance (Fig. 7.30). Therefore, the development of a processing technique to enlarge the contact area is also needed. In the configuration shown in Fig. 7.29(d), the cell does not contain any liquid parts and it is called all-solid-state Li-air battery. This configuration can eliminate the issue of leakage of liquid electrolyte, although the solid electrolyte faces an even more stringent requirement, i.e. high stability against both of Li metal and air. As mentioned above, the solid electrolytes in Li-air battery play an important role to protect Li metal from the aqueous electrolyte/moisture in air (Figs. 7.29(b)–7.29(d)) and to prevent mixing of organic and aqueous electrolytes (Fig. 7.29(b)). In next section, the configurations from Figs. 7.29(b)–7.29(d) that contain the solid electrolytes are described in detail.

7.4.2   *Solid electrolyte for Li-air battery containing aqueous electrolyte*

When aqueous electrolyte is used, the cell reactions in the Li-air battery are different from that in the organic electrolyte

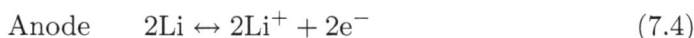

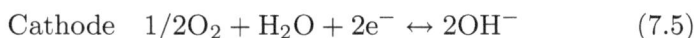

$$\text{Anode} \quad 2\text{Li} \leftrightarrow 2\text{Li}^+ + 2e^- \tag{7.4}$$

$$\text{Cathode} \quad 1/2\text{O}_2 + \text{H}_2\text{O} + 2e^- \leftrightarrow 2\text{OH}^- \tag{7.5}$$

$$\text{Total} \quad 2\text{Li} + 1/2\text{O}_2 + \text{H}_2\text{O} \leftrightarrow 2\text{LiOH} \tag{7.6}$$

In this case, the energy density is calculated to $1910\,\mathrm{Wh\,kg^{-1}}$ with an open circuit voltage of $3.0\,\mathrm{V}$. The open circuit voltage of the aqueous Li-air cell changed depending on the pH value of the aqueous electrolyte [88]. The aqueous electrolyte at the cathode side (catholyte) can dissolve the discharge products, which can improve cycle efficiency, power performance and volumetric energy density of the Li-air batteries [89]. Therefore, this type of Li-air battery has been considered to be a promising configuration among the current techniques for Li-air battery.

Due to high reactivity of Li metal anode with aqueous electrolyte, a protective layer is definitely needed. The protective layer should have high Li-ion conductivity and be stable in the aqueous and organic electrolytes as well as Li metal and so on. Solid electrolytes with high Li-ion conductivity have been expected to serve as a protective layer. The sulfide-based electrolytes can be eliminated as a candidate for the protective layer because of high reactivity with moisture. Therefore, high Li-ion conductive oxide-based solid electrolytes such as perovskite, garnet and NASCION are considered to be suitable. Boulant *et al.* studied the stability of $Li_{0.30}La_{0.57}TiO_3$ perovskite (LLTO) in pure water. Figure 7.31 shows the pH evolution of pure water after immersion of LLTO powder at $70^\circ\mathrm{C}$ [90]. After immersion of LLTO powder into pure water, the pH value increased abruptly. This can be attributed to the exchange reaction of H^+ for Li^+ in LLTO, as demonstrated by the following reaction:

$$Li_{0.30}La_{0.57}TiO_3 + yH_2O \rightarrow (Li_{0.30-y}H_y)La_{0.57}TiO_3 + yLi^+ + yOH^-$$

The exchange reaction was also observed in acidic 2M HNO_3 solution [91]. It was found by ^1H MAS NMR that the exchanged H^+ occupied three different environments. Furthermore, it was reported that the exchange reaction occurred even with moisture in ambient air at room temperature [92]. From these results, it can be concluded that the perovskite LLTO is not stable in aqueous electrolyte by the H^+ exchange reaction. Additionally, by facile reduction of Ti^{4+} into Ti^{3+}, LLTO is not stable in contact with Li metal.

H^+ exchange reaction for Li^+ is also observed in the garnet-type solid electrolyte. Ma *et al.* studied on the H^+ exchange of

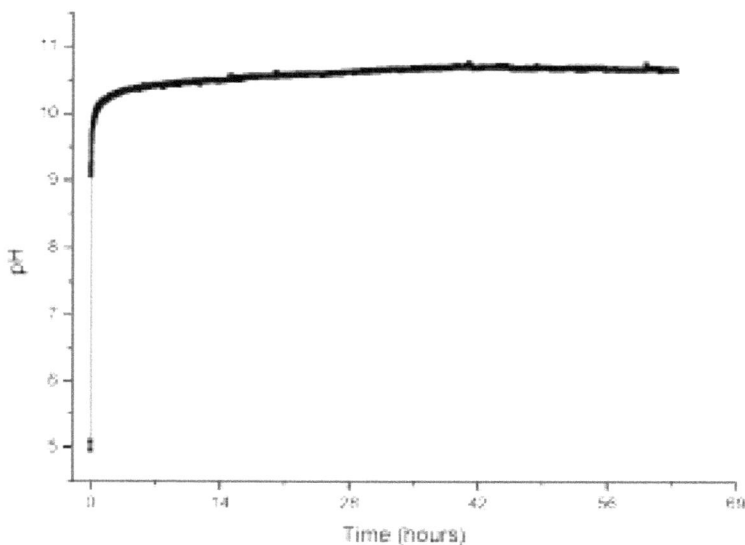

Fig. 7.31   pH evolution of water after immersion of LLTO powder at 70°C.

$Li_7La_3Zr_2O_{12}$ (LLZ) solid electrolyte. The H^+ exchange for Li^+ occurred when LLZ was immersed in de-ionized water [93], but maintained the cubic garnet structure even at the high exchange rate of 63.6%. Also, they found that the H^+ exchange reaction was reversible without any structural change in 2M LiOH solution. The same behavior was observed in other garnets such as $Li_5La_3Ta_2O_{12}$ (LLTa) [94], $Li_5La_3Nb_2O_{12}$ (LLNb) [95], Ta-doped LLZ [96], Ba-doped $Li_5La_3Nb_2O_{12}$ [97], Ba-doped $Li_5La_3Ta_2O_{12}$ [98], Ca-doped $Li_5La_3Nb_2O_{12}$ [99], Sc-doped $Li_5La_3Nb_2O_{12}$, Sc-doped $Li_5La_3Ta_2O_{12}$, Nb-doped LLZ [100] and so on. Figure 7.32 depicts a plot of pH evolution as a function of time for $Li_5La_3Nb_2O_{12}$ in water. The increase of the pH value was observed just after immersion of the solid electrolyte into water like LLT. The stable pH value reached to 12 within 25 s. Also, the stability of LLNb in various concentrations of HCl was studied. XRD spectra of LLNb after immersion in various concentrations of HCl are shown in Fig.7.33. At high concentration, the crystal structure completely decomposed. The crystal structure was retained after immersion in

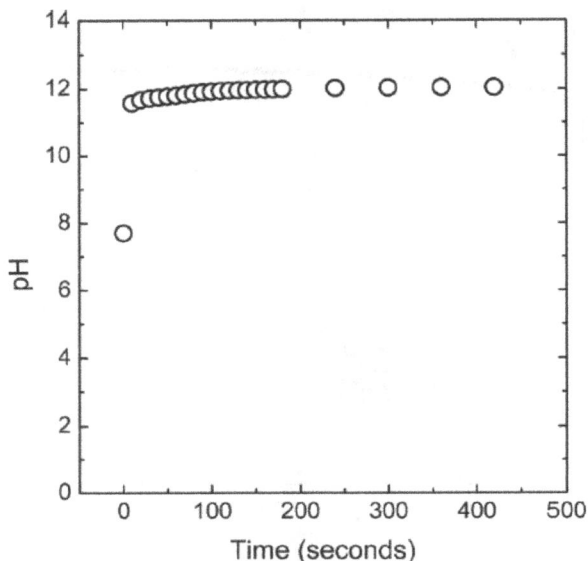

Fig. 7.32   A plot of the pH value of water after immersion of LLNb powder [97].

0.01M HCl [97]. The ion-exchange rate would be influenced by the sintering temperature of the garnet. Truong *et al.* reported that 93% of the Li-ion in $Li_5La_3Nb_2O_{12}$ sintered at 950°C was exchanged after immersion in water for 14 days [101]. On the contrary, Nemori *et al.* observed that only 35% of the Li-ion was exchanged even after 35 days of immersion in the same garnet sintered at 1150°C [100]. The lattice constant of garnet becomes larger by partial substitution of H^+ (Fig. 7.34). A similar behavior was observed for H^+-exchanged garnet by moisture in air. The partial H^+-exchanged garnet formed a low ionic conductive phase that had a similar XRD pattern to the original phase, but with a slightly larger lattice parameter [102]. The garnets are more stable in alkaline solution than in water. No conductivity change was observed in the Ta- and Nb-based garnets after immersion in LiCl + LiOH aqueous solution for 2 weeks at 25°C [100]. Regarding the stability of the garnet solid electrolytes against Li metal, LLZ and LLTa were verified to be stable in contact with Li metal by both calculation [103] and experimental studies [104,105]. LLNb is not stable against Li metal. Nemori *et al.* observed

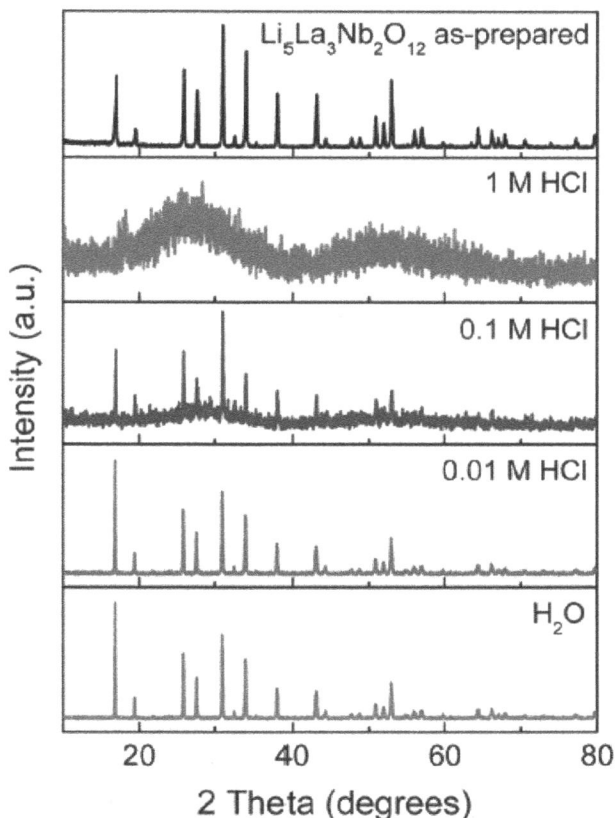

Fig. 7.33 XRD patterns of LLNb after immersion in various concentrations of HCl [97].

a reduction of Zr-doped LLNb by Li metal [100]. However, it was also reported that the interfacial resistance of Li-/Zr-doped LLNb was not increased even after a long duration [106,107]. Although the stability of Nb-based garnet against Li metal is still widely debated, LLZ- and LLTa-based garnet are expected to serve as the protective layer in alkaline aqueous electrolyte.

The most widely studied solid electrolytes for Li-air batteries are NASICON-type solid electrolytes. Especially, $Li_{1+x+y}Ti_{2-x}Al_x P_{3-y}Si_yO_{12}$ (LTAP) glass-ceramics have been reported to be unstable in LiOH aqueous solution, but stable in $LiNO_3$ and $LiCl_3$ aqueous

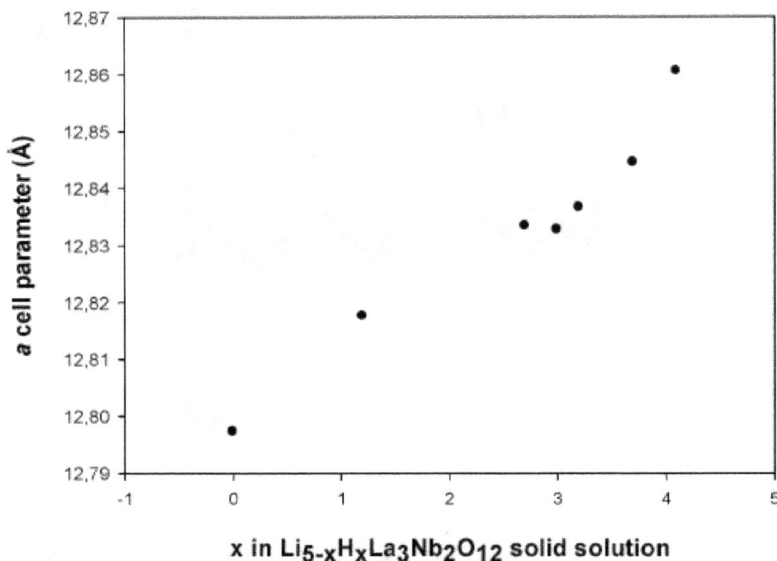

Fig. 7.34   Lattice parameter of H^+-exchanged LLNb [102].

solution [108]. However, the buffer layer between Li metal and LTAP is needed because LTAP contains Ti^{4+} which is easily reduced to Ti^{3+} by Li metal. As the buffer layer, PEO-based polymer electrolyte has been studied [109, 110]. The Li-air cell composed of this three-layer Li anode could be operated for 100 h [110] (Fig. 7.35). In this three-layer Li metal anode, a predominant resistance existed at the interface between Li metal/polymer electrolyte/solid electrolyte [109, 111]. Figure 7.36 shows a complex impedance plot of a Li/PEO/LATP/PEO/Li cell with and without $BaTiO_3$ filler [109]. The $BaTiO_3$ filler improves Li-ion conductivity of the PEO polymer. The large depressed semicircle in the medium frequency range is due to the Li/polymer and polymer/LATP interfaces. These resistances appear at a similar frequency range, and so the large depressed semicircle was observed. It can be understood that these interface resistances are the largest resistance component in the cell. The resistance could be reduced by addition of $BaTiO_3$ filler probably

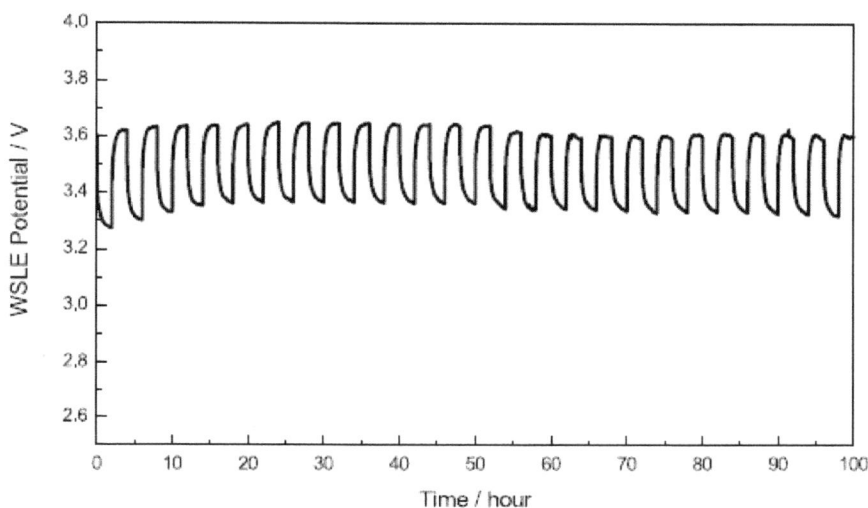

Fig. 7.35   Potential change of Li/PEO/1M LiCl + 0.002 M LiOH/Pt black cell [110].

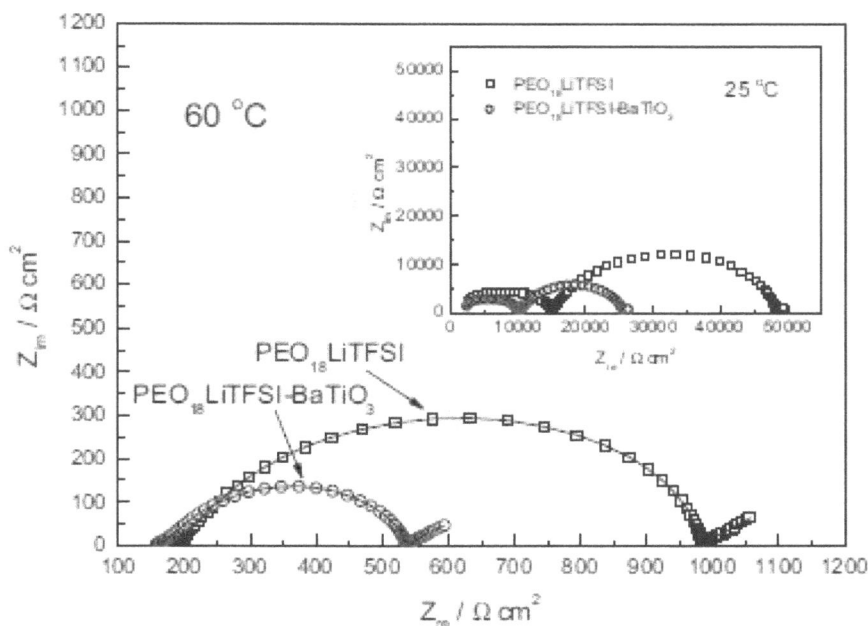

Fig. 7.36   A complex impedance plot of Li/PEO/LATP/PEO/Li cell with and without BaTiO₃ filler.

due to an increase in the softness of the PEO caused by the addition of the $BaTiO_3$ filler. Also, the Li-air cell using the three-layer anode and mixed aqueous solution of CH_3COOH and CH_3COOLi was fabricated. The cell reaction can be described as follows [112]:

Anode $2Li \leftrightarrow 2Li^+ + 2e^-$

Cathode $0.5O_2 + 2CH_3COOH + 2e^- \leftrightarrow H_2O + 2CH_3COO^-$

Total $2Li + 0.5O_2 + 2CH_3COOH \leftrightarrow 2CH_3COOLi + H_2O$

A schematic diagram of the cell is shown in Fig. 7.37, and charge and discharge curves are depicted in Fig. 7.38. The cell shows extremely high energy density of $779\,Wh\,kg^{-1}$, which was twice as high as that of the commercial graphite/$LiCoO_2$ cell. When compared with LTAP,

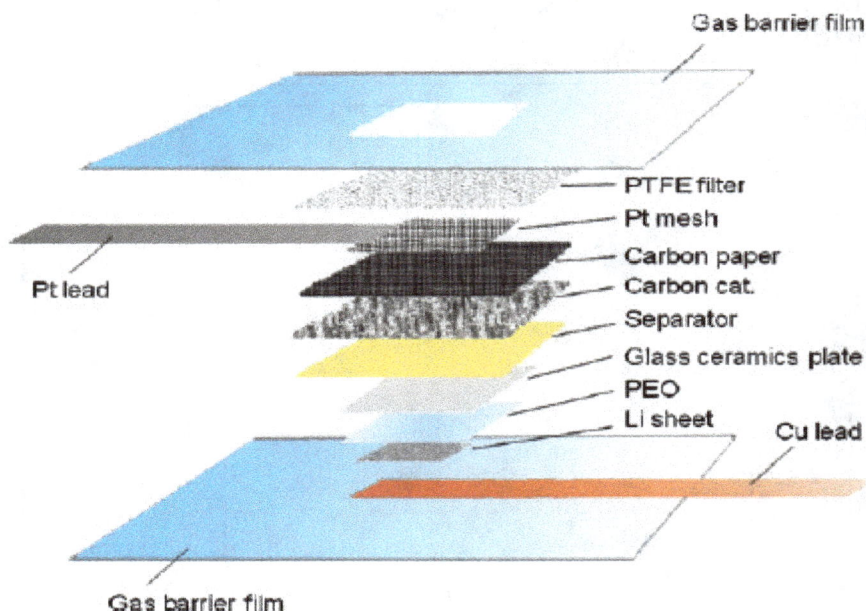

Fig. 7.37   Schematic diagram of Li-air cell.

Source: Reproduced from Ref. [112] with permission from Royal Society of Chemistry.

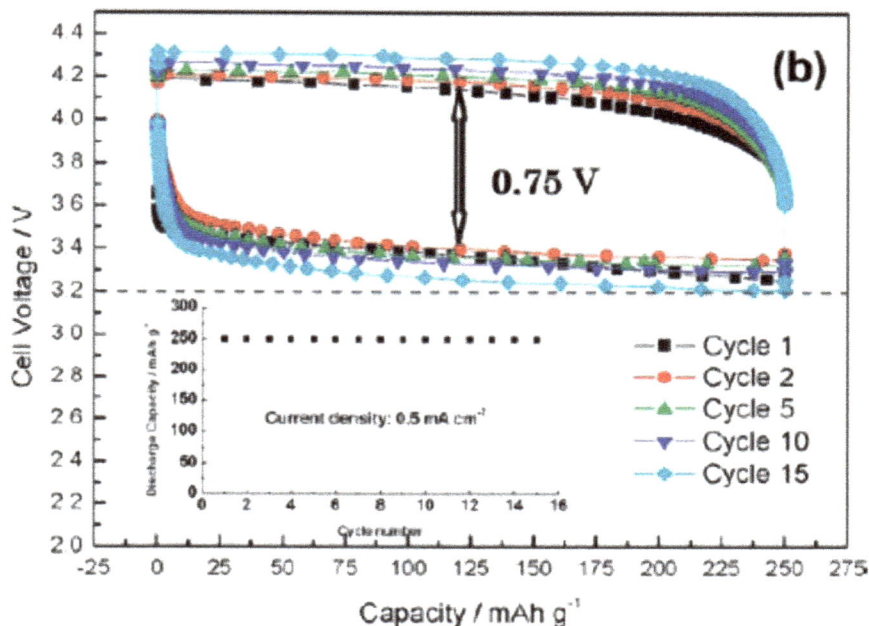

Fig. 7.38   Charge and discharge curves of Li-air cell depicted in Fig. 7.37.
Source: Reproduced from Ref. [112] with permission from Royal Society of Chemistry.

other NASICON-type solid electrolytes have seldom been studied. Zhang *et al.* prepared the three-layer Li anode using tape casting technique of $Li_{1.4}Al_{0.4}Ge_{1.6}(PO_4)_3$ (LAGP) solid electrolyte [113]. By addition of nano-TiO_2 powder, the electrical conductivity of LAGP thin layer was increased. Although LAGP possesses high degree of kinetic stability in contact with Li metal, TiO_2 addition makes the thin layer unstable against Li metal. The cell using Li/PEO/TiO_2-LAGP electrode and saturated LiCl aqueous solution was operated for 20 h at $0.5\,mA\,cm^{-2}$, and no degradation was observed. Adams *et al.* constructed cells of the type demonstrated in Fig. 7.29(b) using LAGP, $LiPF_6$ dissolved in EC-DMC as the organic electrolyte and various aqueous electrolytes, and examined the battery performance of the cells [114]. The performance was influenced by the aqueous

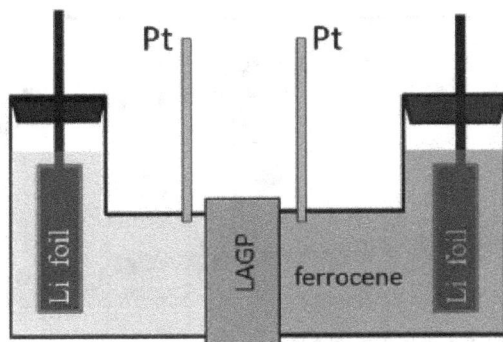

Fig. 7.39   Cell configuration for the cross-over test.

electrolytes. The highest performance was obtained in acidified LiCl (pH = 2–3).

The protective layer should be made of pore-free and well-sintered ceramics to prevent penetration of the aqueous electrolyte into the Li metal anode. To check the sinterability of solid electrolyte, a comparison of the actual density of the sintered solid electrolyte against the theoretical one has been widely used. The density of the solid electrolyte is usually calculated by the Archimedes method. As another method to check the penetration of liquid electrolyte, electrochemical cross-over test has been employed (Fig. 7.39) [115]. The cell for the cross-over test is composed of two chambers that are filled with liquid electrolyte. Electrochemically active ferrocene $(Fe(C_5H_5)_2)$ is added into one chamber. After Pt and Li metal electrodes are inserted into both chambers, a redox reaction of ferrocene in both chambers is monitored by cyclic voltammetry (Fig. 7.40). If the redox reaction of ferrocene is not observed in the chamber where ferrocene is not added, it is proved that the cross-over does not occur.

7.4.3   *Solid electrolytes for all-solid-state Li-air batteries*

The all-solid-state Li-air battery can eliminate any volatile and flammable liquid electrolyte and is recognized as the most feasible

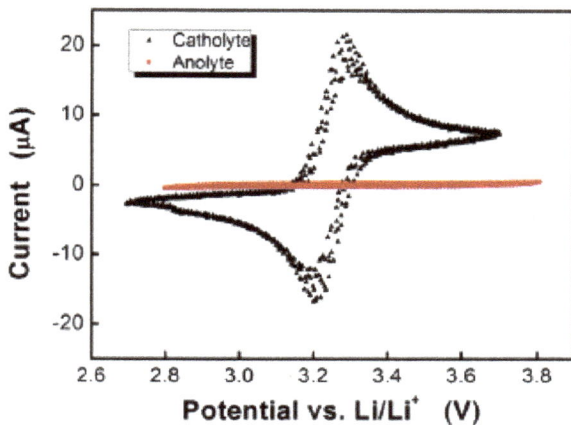

Fig. 7.40 CV curves for the cross-over test.

configuration for safety and long life [116]. The cell reactions of all-solid-state Li-air battery can be described as being similar to those in Eqs. (7.1)–(7.3). Li_2O and Li_2O_2 are formed on the cathode during discharge. Abraham *et al.* first reported all-solid-state Li-air battery using polymer electrolyte [117]. The battery exhibited high capacity, about $1200\,mA\,h\,g^{-1}$ basis of carbon which was added to provide electrical conductivity to the cathode (Fig. 7.41). This value was nearly one order higher than that of the $LiCoO_2$ cathode. This result was the green signal for the development of the all-solid-state Li-air cell. However, the cell demonstrated poor cyclability. After that, Kumar *et al.* prepared all-solid-state Li-air battery using LAGP solid electrolyte. The cell was successfully operated at 0.05–$0.1\,mA\,cm^{-2}$ at 75–95°C and OCV was changed with oxygen partial pressure [118]. Figure 7.42 depicts the first two charge and discharge cycles of the all-solid-state Li-air battery at 95°C. They did not mention a reason for the higher capacity of the second cycle than the first cycle. From the charge and discharge curves, it can be seen that the cell still has a cyclability problem. Moreover, the discharge capacity was only $56\,mA\,h\,g^{-1}$, which was too low when compared with the Li-air cell with liquid electrolyte. The low capacity could be explained by the low reaction area at the air cathode. The reaction at the air cathode

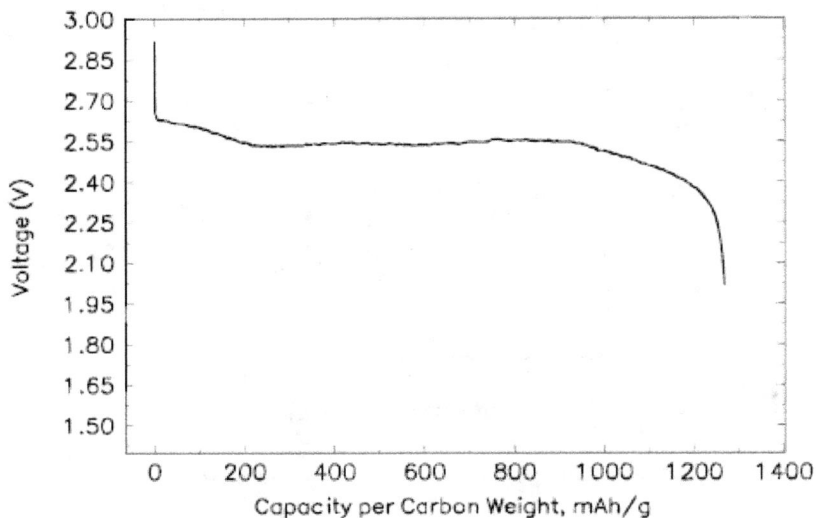

Fig. 7.41　Discharge curve of all-solid-state Li-air cell [117].

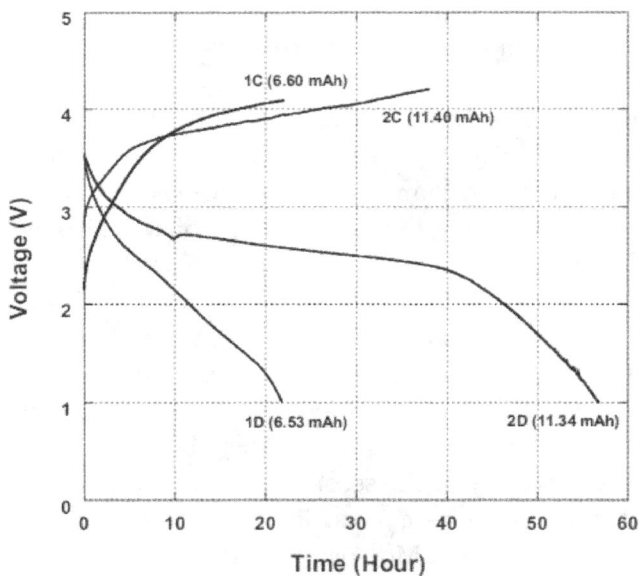

Fig. 7.42　First two charge and discharge curves of the all-solid-state Li-air battery [118].

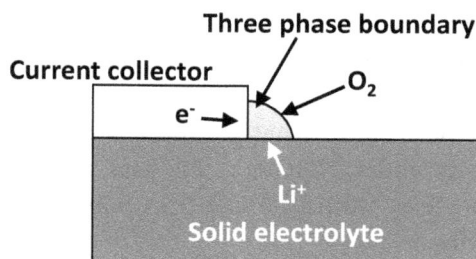

Fig. 7.43 Schematic image of the three-phase boundary.

should proceed at the three-phase boundary, i.e. air, electrolyte and current collector (Fig. 7.43). To improve the capacity, the reaction area should be increased significantly. For this purpose, Pt patterned electrode was developed [119]. The patterned electrode can provide a large contact area between the air and electrode. Additionally, Pt can catalyze both oxygen reduction and oxygen revolution reactions. The cell composed of Li/polymer electrolyte/LATP/Pt pattern electrode showed a discharge capacity of $1750\,\mathrm{mA\,h\,g^{-1}}$. The development of a suitable configuration of the cathode is also key for the all-solid-state Li-air cell.

7.4.4 *Summary*

The Li-air batteries have attracted a lot of attention in recent years due to their extremely high energy density, which is comparable to that of the present gasoline system. However, performance in terms of long-term stability and power density (operation under high current condition) does not satisfy the requirement for practical application. One of the reasons for this is the formation of insulative Li_2CO_3 at cathode when air is used. Therefore, novel structures of the cathode must be studied. The essential requirement of the solid electrolyte in the Li-air battery is high stability against Li metal and liquid electrolyte or air as well as high Li-ion conductivity. The Li-ion conductivity and resistance at interfaces between Li-metal/solid electrolyte and liquid electrolyte/solid electrolyte directly relate to the power density of the cell. In addition, the stability of the interfaces concerns long-term operation. These problems must be solved to enable its practical use. Some breakthrough is needed, but it seems some more time is still needed to solve the problems. Therefore, it can be assumed that its application for operation under low current density and with a short life time, such as a replacement of Li primary battery, would be the first market of application for the Li-air batteries.

References

[1] J. N. Mrgudich: Conductivity of silver iodide pellets for solid — electrolyte batteries, *J. Electrochem. Soc.* 107 (1960) 475–479.

[2] B. B. Owens, P. M. Skarstad: Ambient temperature solid state batteries, *Solid State Ionics* 53–56 (1992) 665–672.

[3] V. Thangadurai, W. Weppner: Investigations on electrical conductivity and chemical compatibility between fast lithium ion conducting garnet-like $Li_6BaLa_2Ta_2O_{12}$ and lithium battery cathodes, *J. Power Sources* 142 (2005) 339–344.

[4] N. J. Dudney: Solid-state thin-film rechargeable batteries, *Mater. Sci. Eng. B* 116 (2005) 245–249.

[5] B. Wang, J. B. Bates, F. X. Hart, B. C. Sales, R. A. Zuhr, J. D. Robertson: Characterization of thin-film rechargeable lithium batteries with lithium cobalt oxide cathodes, *J. Electrochem. Soc.* 143 (1996) 3203–3213.

[6] F. Le Cras, B. Pecquenard, V. Dubios, V.-P. Phan, D. Guy-Bouyssou: All-solid-state lithium-ion microbatteries using silicon nanofilm anodes: high performance and memory effect, *Adv. Energy Mater.* 5 (2015) 1501061.

[7] S. Tintignac, R. Baddour-Hadjean, J. P. Pereira-Ramos, R. Salot: High rate bias sputtered $LiCoO_2$ thin films as positive electrode for all-solid-state lithium microbatteries, *Electrochim. Acta* 146 (2014) 472–476.

[8] F. Huang, Z.-W. Fu, Q.-Z. Qin: A novel $Li_2Ag_{0.5}V_2O_5$ composite film cathode for all-solid-state lithium batteries, *Electrochem. Comm.* 5 (2003) 262–266.

[9] M. Kotobuki, K. Kanamura: Fabrication of all-solid-state battery using $Li_5La_3Ta_2O_{12}$ ceramic electrolyte, *Ceram. Intl.* 39 (2013) 6481–6487.

[10] B. J. Neudecker, N. J. Dudney, J. B. Bates: "Lithium-free" thin-film battery with *in situ* plated Li anode, *J. Electrochem. Soc.* 147 (2000) 517–523.

[11] F. Kirino, Y. Ito, K. Miyauchi, T. Kudo: Electrochemical behavior of amorphous thin films of sputtered V_2O_5-WO_3 mixed conductors, *Nippon Kagakukaishi* 1986 (1986) 445–450.

[12] K. Kanehori, K. Matsumoto, K. Miyauchi, T. Kudo: Thin film solid electrolyte and its application to secondary lithium cell, *Solid State Ionics* 9–10 (1983) 1445–1448.

[13] K. Kanehori, Y. Ito, F. Kirino, K. Miyauchi, T. Kudo: Titanium disulfide films fabricated by plasma CVD, *Solid State Ionics* 18–19 (1986) 818–822.

[14] S. D. Jones, J. R. Akridge: A thin film solid state microbattery, *Solid State Ionics* 53–56 (1992) 628–634.

[15] http://www.cymbet.com/

[16] M. Baba, N. Kumagai, H. Kobayashi, O. Nakano, K. Nishidate: Fabrication and electrochemical characteristics of all-solid-state lithium-ion batteries using V_2O_5 thin films for both electrodes, *Electrochem. Solid State Lett.* 2 (1999) 320–322.

[17] N. J. Dudney, J. B. Bates, R. A. Zuhr, S. Young, J. D. Robertson, H. P. Jun, S. A. Hackney: Nanocrystalline $Li_xMn_{2-y}O_4$ cathodes for solid-state thin-film rechargeable lithium batteries, *J. Electrochem. Soc.* 146 (1999) 2455–2464.

[18] J. B. Bates, D. Lubben, N. J. Dudney, F. X. Hart: 5 Volt Plateau in $LiMn_2O_4$ thin films, *J. Electrochem. Soc.* 142 (1995) L149–L151.

[19] Y. S. Park, S. H. Lee, B. I. Lee, S. K. Joo: All-solid-state lithium thin-film rechargeable battery with lithium manganese oxide, *Electrochem. Solid State Lett.* 2 (1999) 58–59.

[20] B. J. Neudecker, R. A. Zuhr, J. D. Robertson, J. B. Bates: Lithium manganese nickel oxides $Li_x(Mn_yNi_{1-y})_{2-x}O_2$: I. Synthesis and characterization of thin films and bulk phases, *J. Electrochem. Soc.* 145 (1998) 4148–4159.

[21] B. J. Neudecker, R. A. Zuhr, J. D. Robertson, J. B. Bates: Lithium manganese nickel oxides $Li_x(Mn_yNi_{1-y})_{2-x}O_2$: II. Electrochemical studies on thin-film batteries, *J. Electrochem. Soc.* 145 (1998) 4160–4168.

[22] J. B. Bates, N. J. Dudney, D. C. Lubben, G. R. Gruzalski, B. S. Kwak, X. Yu, R. A. Zuhr: Thin-film rechargeable lithium batteries, *J. Power Sources* 54 (1995) 58–62.

[23] J. B. Bates, N. J. Dudney, B. Neudecker, A. Ueda, C. D. Evans: Thin-film lithium and lithium-ion batteries, *Solid State Ionics* 135 (2000) 33–45.

[24] B. J. Neudecker, R. A. Zuhr, J. B. Bates: Lithium silicon tin oxynitride (Li_ySiTON): high-performance anode in thin-film lithium-ion batteries for microelectronics, *J. Power Sources* 81–82 (1999) 27–32.

[25] J. Tan, A. Tiwari: Fabrication and characterization of $Li_7La_3Zr_2O_{12}$ thin films for lithium ion battery, *ECS Solid State Lett.* 1 (2012) Q57–Q60.

[26] D. J. Kalita, S. H. Lee, K. S. Lee, D. H. Ko, Y. S. Yoon: Ionic conductivity properties of amorphous Li-La-Zr-O solid electrolyte for thin film batteries, *Solid State Ionics* 229 (2012) 14–19.

[27] C.-W. Ahn, J.-J. Choi, J. Ryu, B.-D. Hahn, J.-W. Kim, W.-H. Yoon, J.-H. Choi, D.-S. Park: Microstructure and ionic conductivity in $Li_7La_3Zr_2O_{12}$ film prepared by aerosol deposition method, *J. Electrochem. Soc.* 162(1) (2015) A60–A63.

[28] R.-J. Chen, M. Huang, W.-Z. Huang, Y. Shen, Y.-H. Lin, C.-W. Nan: Sol–gel derived Li-La-Zr-O thin films as solid electrolytes for lithium-ion batteries, *J. Mater. Chem. A* 2(33) (2014) 13277–13282.

[29] G. Tan, F. Wu, L. Li, Y. Liu, R. Chen: Magnetron sputtering preparation of nitrogen-incorporated lithium-aluminum-titanium phosphate based thin

film electrolytes for all-solid-state lithium ion batteries, *J. Phys. Chem. C* 116 (2012) 3817–3826.

[30]   M. Morcrette, A. Gutierrez-Llorente, A. Laurent, J. Perrière, P. Barboux, J. P. Boilot, O. Raymond, T. Brousse: Growth by laser ablation of Ti-based oxide films with different valency states, *Appl. Phys. A: Mater. Sci. Process.* 67(4) (1998) 425–428.

[31]   S. Furusawa, H. Tabuchi, T. Sugiyama, S. Tao, J. TS. Irvine: Ionic conductivity of amorphous lithium lanthanum titanate thin film, *Solid State Ionics* 176(5) (2005) 553–558.

[32]   J. K. Ahn, S. G. Yoon: Characteristics of amorphous lithium lanthanum titanate electrolyte thin films grown by PLD for use in rechargeable lithium microbatteries, *Electrochem. Solid-State Lett.* 8(2) (2005) A75–A78.

[33]   J. K. Ahn, S. G. Yoon: Characteristics of perovskite ($Li_{0.5}La_{0.5}$) TiO_3 solid electrolyte thin films grown by pulsed laser deposition for rechargeable lithium microbattery, *Electrochimica Acta* 50(2) (2004) 371–374.

[34]   M. Morales, P. Laffez, D. Chateigner, I. Vickridge: Characterisation of lanthanum lithium titanate thin films deposited by radio frequency sputtering on [100]-oriented MgO substrates, *Thin Solid Films* 418(2) (2002) 119–128.

[35]   C. Li, B. Zhang, Z. Fu: Physical and electrochemical characterization of amorphous lithium lanthanum titanate solid electrolyte thin-film fabricated by e-beam evaporation, *Thin Solid Films* 515(4) (2006) 1886–1892.

[36]   Z. Zheng, H. Fang, F. Yang, Z. Liu, Y. Wang: Amorphous LiLaTiO3 as solid electrolyte material, *J. Electrochem. Soc.* 161(4) (2014) A473–A479.

[37]   K. Kitaoka, H. Kozuka, T. Hashimoto, T. Yoko: Preparation of $Li_{0.5}La_{0.5}TiO_3$ perovskite thin films by the sol–gel method, *J. Mater. Sci.* 32(8) (1997) 2063–2070.

[38]   T. Aaltonen, M. Alnes, O. Nilsen, L. Costelle, H. Fjellvåg: Lanthanum titanate and lithium lanthanum titanate thin films grown by atomic layer deposition, *J.f Mater. Chem.* 20(14) (2010) 2877–2881.

[39]   J. Z. Lee, Z. Wang, HL. Xin, TA. Wynn, Y. S. Meng: Amorphous lithium lanthanum titanate for solid-state microbatteries, *J. Electrochem. Soc.* 164(1) (2017) A6268–A6273.

[40]   Y. Ito, A. Sakuda, T. Ohtomo, A. Hayashi, M. Tatsumisago: Preparation of Li_2S-GeS_2 solid electrolyte thin films using pulsed laser deposition, *Solid State Ionics* 236 (2013) 1–4.

[41]   A. Sakuda, A. Hayashi, S. Hama, M. Tatsumisago: Preparation of highly lithium-ion conductive $80Li_2S \cdot 20P_2S_5$ thin-film electrolytes using pulsed laser deposition, *J. Am. Ceram. Soc.* 93 (2010) 765–768.

[42]   Y. Ito, A. Sakuda, T. Ohtomo, A. Hayashi, M. Tatsumisago: Li_4GeS_4-Li_3PS_4 electrolyte thin films with highly ion-conductive crystals prepared by pulsed laser deposition, *J. Ceram. Soc. Jpn.* 122 (2014) 341–345.

[43]   A. Aboulaich, R. Bouchet, G. Delaizir, V. Seznec, L. Tortet, M. Morcrette, P. Rozier, J.-M. Tarascon, V. Viallet, M. Dolle: A new approach to develop safe all-inorganic monolithic Li-ion batteries, *Adv. Energy Mater.* 1(2) (2011) 179–183.

[44] M. Kotobuki, Y. Suzuki, H. Munakata, K. Kanamura, Y. Sato, K. Yamamoto, T. Yoshida: Fabrication of three-dimensional battery using ceramic electrolyte with honeycomb structure by sol-gel process, *J. Electrochem. Soc.* 157(4) (2010) A493–A498.

[45] M. Kotobuki, Y. Suzuki, K. Kanamura, Y. Sato, K. Yamamoto, T. Yoshida: A novel structure of ceramics electrolyte for future lithium battery, *J. Power Sources* 196 (2011) 9815–9819.

[46] M. Kotobuki, H. Munakata, K. Kanamura: Fabrication of all-solid-state rechargeable lithium-ion battery using mille-feuille structure of $Li_{0.35}La_{0.55}TiO_3$, *J. Power Sources* 196 (2011) 6947–6950.

[47] M. Kotobuki, Y. Suzuki, H. Munakata, K. Kanamura, Y. Sato, K. Yamamoto, T. Yoshida: Compatibility of $LiCoO_2$ and $LiMn_2O_4$ cathode materials for $Li_{0.55}La_{0.35}TiO_3$ electrolyte to fabricate all-solid-state lithium battery, *J. Power Sources* 195 (2010) 5784–5788.

[48] K. Nagata, T. Nanno: All solid battery with phosphate compounds made through sintering process, *J. Power Sources* 174(2) (2007) 832–837.

[49] E. Kobayashi, L. S. Plashnitsa, T. Doi, S. Okada, J.-I. Yamaki: Electrochemical properties of Li symmetric solid-state cell with NASICON-type solid electrolyte and electrodes, *Electrochem. Commun.* 12(7) (2010) 894–896.

[50] G. Delaizir, V. Viallet, A. Aboulaich, R. Bouchet, L. Tortet, V. Seznec, M. Morcrette, J.-M. Tarascon, P. Rozier, M. Dolle: The stone age revisited: building a monolithic inorganic lithium-ion battery, *Adv. Funct. Mater.* 22(10) (2012) 2140–2147.

[51] K. Hoshina, K. Dokko, K. Kanamura: Investigation on electrochemical interface between $Li_4Ti_5O_{12}$ and $Li_{1+x}Al_xTi_{2-x}(PO_4)_3$ NASICON type solid electrolyte, *J. Electrochem. Soc.* 152(11) (2005) A2138–A2142.

[52] K. Dokko, K. Hoshina, H. Nakano, K. Kanamura: Preparation of $LiMn_2O_4$ thin-film electrode on $Li_{1+x}Al_xTi_{2-x}(PO_4)_3$ NASICON-type solid electrolyte, *J. Power Sources* 174(2) (2007) 1100–1103.

[53] J. Xie, N. Imanishi, T. Zhang, A. Hirano, Y. Takeda, O. Yamamoto: Li-ion transport in all-solid-state lithium batteries with $LiCoO_2$ using NASICON-type glass ceramic electrolytes, *J. Power Sources* 189(1) (2009) 365–370.

[54] M. Kotobuki, Y. Suzuki, H. Munakata, K. Kanamura, Y. Sato, K. Yamamoto, T. Yoshida: Effect of sol composition on solid electrode/solid electrolyte interface for all-solid-state lithium ion battery, *Electrochim. Acta* 56 (2011) 1023–1029.

[55] K. Takada, S. Kondo: Lithium ion conductive glass and its application to solid state batteries, *Ionics* 4 (1998) 42–47.

[56] Y. Yin, S. Xin, Y. Guo, L. Wan: Lithium–sulfur batteries: Electrochemistry, materials, and prospects, *Angew. Chem. Int. Ed.* 52 (2013) 13186–13200.

[57] S. Yubuchi, Y. Ito, T. Matsuyama, A. Hayashi, M. Tatsumisago: 5V class $LiNi_{0.5}Mn_{1.5}O_4$ positive electrode coated with Li_3PO_4 thin film for all-solid-state batteries using sulfide solid electrolyte, *Solid State Ionics* 285 (2016) 79–82.

[58] N. Ohta, K. Takada, L. Q. Zhang, R. Z. Ma, M. Osada, T. Sasaki: Enhancement of the high-rate capability of solid-state lithium batteries by nanoscale interfacial modification, *Adv. Mater.* 18 (2006) 2226–2229.

[59] H. Kitaura, A. Hayashi, K. Tadanaga, M. Tatsumisago: Electrochemical performance of all-solid-state lithium secondary batteries with Li-Ni-Co-Mn oxide positive electrodes, *Electrochim. Acta* 55 (2010) 8821–8828.

[60] H. Kitaura, A. Hayashi, K. Tadanaga, M. Tatsumisago: Improvement of electrochemical performance of all-solid-state lithium secondary batteries by surface modification of $LiMn_2O_4$ positive electrode, *Solid State Ionics* 192 (2011) 304–307.

[61] Y. Seino, T. Ota, K. Takada: High rate capabilities of all-solid-state lithium secondary batteries using $Li_4Ti_5O_{12}$-coated $LiNi_{0.8}Co_{0.15}Al_{0.05}O_2$ and a sulfide-based solid electrolyte, *J. Power Sources* 196 (2011) 6488–6492.

[62] N. Ohta, K. Takada, I. Sakaguchi, L. Q. Zhang, R. Z. Ma, K. Fukuda, M. Osada, T. Sasaki: $LiNbO_3$-coated $LiCoO_2$ as cathode material for all solid-state lithium secondary batteries, *Electrochem. Comm.* 9 (2007) 1486–1490.

[63] K. Takada, N. Ohta, L. Q. Zhang, K. Fukuda, I. Sakaguchi, R. Z. Ma, M. Osada, T. Sasaki: Interfacial modification for high-power solid-state lithium batteries, *Solid State Ionics* 179 (2008) 1333–1337.

[64] X. Xu, K. Takada, K. Watanabe, I. Sakaguchi, K. Akatsuka, B. T. Hang, T. Ohnishi, T. Sasaki: Self-organized core-shell structure for high-power electrode in solid-state lithium batteries, *Chem. Mater.* 23 (2011) 3798–3804.

[65] A. Sakuda, A. Hayashi, M. Tatsumisago: Interfacial observation between $LiCoO_2$ electrode and Li_2S-P_2S_5 solid electrolytes of all-solid-state lithium secondary batteries using transmission electron microscopy, *Chem. Mater.* 22 (2010) 949–956.

[66] S. Evers, F. Nazar: New approaches for high energy density lithium-sulfur battery cathodes, *Acc. Chem. Res.* 46 (2013) 1135–1143.

[67] A. Manthiram, Y. Fu, S. Chung, C. Zu, Y. Su: Rechargeable lithium-sulfur batteries, *Chem. Rev.* 114 (2014) 11751–11787.

[68] S. Xin, L. Gu, N. Zhao, Y. Yin, L. Zhou, Y. Gu, L. Wan: Smaller sulfur molecules promise better lithium–sulfur batteries, *J. Am. Chem. Soc.* 134 (2012) 18510–18513.

[69] X. Ji, K. Lee, F. Nazar: A highly ordered nanostructured carbon-sulphur cathode for lithium-sulphur batteries, *Nat. Mater.* 8 (2009) 500–506.

[70] Y. Yang, G. Yu, J. Cha, H. Wu, M. Vosgueritchian, Y. Yao, Z. Bao, Y. Cui: Improving the performance of lithium-sulfur batteries by conductive polymer coating, *ACS Nano* 5 (2011) 9187–9193.

[71] N. Jayaprakash, J. Shen, S. Moganty, A. Corona, L. Archer: Porous hollow carbon@sulfur composites for high-power lithium-sulfur batteries, *Angew. Chem. Int. Ed.* 50 (2011) 5904–5908.

[72] H. Li, X. Yang, X. Wang, M. Liu, F. Ye, J. Wang, Y. Qiu, W. Li, Y. Zhang: Dense integration of graphene and sulfur through the soft approach for compact lithium/sulfur battery cathode, *Nano Energy* 12 (2015) 468–475.

[73] M. Nagao, K. Suzuki, Y. Imade, M. Tateishi, R. Watanabe, T. Yokoi, M. Hirayama, T. Tatsumi, R. Kanno: All-solid-state lithium sulfur batteries with three-dimensional mesoporous electrode structures, *J. Power Sources* 330 (2016) 120–126.

[74] H. Nagata, Y. Chikusa: An all-solid-state lithium sulfur battery using two solid electrolytes having different functions, *J. Power Sources* 329 (2016) 268–272.

[75] N. Kamaya, K. Homma, Y. Yamakawa, M. Hirayama, R. Kanno, M. Yonemura, T. Kamiyama, Y. Kato, S. Hama, K. Kawamoto, A. Mitsui: A lithium superionic conductor, *Nat. Mater.* 10 (2011) 682–686.

[76] T. Yamada, S. Ito, R. Omoda, T. Watanabe, Y. Aihara, M. Agostini, U. Ulissi, J. Hassoun, B. Scrosati: All solid-state lithium–sulfur battery using a glass-type P_2S_5–Li_2S electrolyte: benefits on anode kinetics, *J. Electrochem. Soc.* 162 (2015) A646–A651.

[77] F. Han, J. Yue, X. Fan, T. Gao, C. Luo, Z. Ma, L. N. Suo, C. Wang: High-performance all-solid-state lithium-sulfur battery enabled by a mixed-conductive Li_2S nanocomposite, *Nano Lett.* 16 (2016) 4521–4527.

[78] https://en.wikipedia.org/wiki/Metal%E2%80%93air_electrochemical_cell.

[79] H. S. Jadhav, R. S. Kalubarme, A. H. Jadhav, J. G. Seo: Highly stable bilayer LiPON and B_2O_3 added $Li_{1.5}Al_{0.5}Ge_{1.5}(PO_4)_3$ solid electrolytes for non-aqueous rechargeabkle Li-O_2 batteries, *Electrochimica Acta* 199 (2016) 126–132.

[80] Y. Sun: Lithium ion conducting membranes for lithium-air batteries, *Nano Energy* 2 (2013) 801–806.

[81] K. M. Abraham, Z. Jiang: A polymer electrolyte-based rechargeable lithium/oxygen battery, *J. Electrochem. Soc.* 143 (1996) 1–5.

[82] M. Zhang, K. Takahashi, I. Uechi, Y. Takeda, O. Yamamoto, D. Im, D.-J. Lee, B. Chi, J. Pu, J. Li, N. Imanishi: Water-stable lithium anode with $Li_{1.4}Al_{0.4}Ge_{1.6}(PO_4)_3$-$TiO_2$ sheet prepared by tape casting method for lithium-air batteries, *J. Power Sources* 235 (2013) 117–121.

[83] T. Ogasawara, A. Debart, M. Holzapfel, P. Novak, P. G. Bruce: Rechargeable Li_2O_2 electrode for lithium batteries, *J. Am. Chem. Soc.* 128 (2006) 1390–1393.

[84] A. Debart, A. J. Paterson, J. Bao, P. G. Bruce: α-MnO2 Nanowires: A catalysts for the O_2 electrode in rechargeable lithium batteries, *Angew. Chem. Int. Ed.* 47 (2008) 4521–4524.

[85] J. Read, K. Mutolo, M. Ervin, W. Behl, J. Wolfenstine, A. Driedger, D. Foster: Oxygen transport properties of organic electrolytes and performance of lithium/oxygen battery, *J. Electrochem. Soc.* 150(10) (2003) A1351–A1356.

[86] S. S. Sandhu, J. P. Fellner, G. W. Brutchen: Diffusion-limited model for a lithium/air battery with an organic electrolyte, *J. Power Sources* 164 (2007) 365–371.

[87] M. Armand, J.-M. Tarascon: Building better batteries, *Nature* 451 (2008) 652–657.

[88] T. Zhang, S. Liu, N. Imanishi, A. Hirano, Y. Takeda, O. Yamamoto: Water-stable lithium electrode and its application in aqueous lithium/air secondary batteries, *Electrochemistry* 78(5) (2010) 360–362.

[89] A. Manthriam, L. Li: Hybrid and aqueous lithium-air batteries, *Adv. Energy Mater.* 5 (2015) 1401302.

[90] A. Boulant, P. Mury, J. Emery, J.-Y. Buzare, O. Bohnke: Efficient ion exchange of H^+ for Li^+ in $(Li_{0.30}La_{0.57}\square_{0.13})TiO_3$ perovskite in water: Protons as a probe for Li location, *Chem. Mater.* 21 (2009) 2209–2217.

[91] N. S. P. Bhuvanesh, O. Bohnke, H. Duroy, M. P. Crosnier-Lopez, J. Emery, J. L. Fourquet: Topotactic H^+/Li^+ ion exchange on $La_{2/3-x}Li_{3x}TiO_3$: new metastable perovskite phases $La_{2/3-x}TiO_{3-3x}(OH)_{3x}$ and $La_{2/3-x}TiO_{3-3x/2}$ obtained by further dehydration, *Mater. Res. Bull.* 33 (1998) 1681–1691.

[92] A. Boulant, J. F. Bardeau, A. Jouanneaux, J. Emery, J.-Y. Buzare, O. Bohnke: Reaction mechanisms of $Li_{0.30}La_{0.57}TiO_3$ powder with ambient air: H^+/Li^+ exchange with water and Li_2CO_3 formation, *Dalton Trans.* 39 (2010) 3968–3975.

[93] C. Ma, E. Rangasamy, C. Liang, J. Sakamoto, K. L. More, M. Chi: Excellent stability of a lithium-ion-conducting solid electrolyte upon reversible Li+/H+ exchange in aqueous solutions, *Angew. Chem. Int. Ed.* 54 (2015) 129–133.

[94] W. Wang, Q. Fang, G. Hao: Reaction mechanisms of $Li_5La_3Ta_2O_{12}$ powder with ambient air: H^+/Li^+ exchange with water, *Adv. Mater. Res.* 463–464 (2012) 123–127.

[95] L. Truong, V. Thangadurai: First total H^+/Li^+ ion Exchange in garnet-type $Li_5La_3Nb_2O_{12}$ using organic acids and studied on the effect of Li stuffing, *Inorg. Chem.* 51(3) (2012) 1222–1224.

[96] Y. Li, J.-T. Han, S. C. Vogel, C.-A. Wang: The reaction of $Li_{6.5}La_3Zr_{1.5}Ta_{0.5}O_{12}$ with water, *Solid State Ionics* 269 (2015) 57–61.

[97] L. Truong, V. Thangadurai: Soft-chemistry of garnet-type $Li_{5+x}Ba_xLa_{3-x}Nb_2O_{12}$ ($x = 0$, 0.5, 1): Reversible $H^+ \leftrightarrow Li^+$ ion-exchange reaction and their X-ray, ^7Li-MAS NMR, IR and AC impedance spectroscopy characterization, *Chem. Mater.* 23(17) (2011) 3970–3977.

[98] L. Truong, J. Colter, V. Thangadurai: Chemical stability of Li-stuffed garnet-type $Li_{5+x}Ba_xLa_{3-x}Ta_2O_{12}$ ($x = 0$, 0.5, 1) in water: A comparative analysis with the Nb analogue, *Solid State Ionics* 247–248 (2013) 1–7.

[99] C. Galven, E. Suard, D. Mounier, M.-P. Crosnier-Lopez, F. Le Berre: Structural characterization of a new acentric protonated garnet: $Li_{6-x}H_xCaLa_2Nb_2O_{12}$, *J. Mater. Res.* 28(16) (2013) 2147–2153.

[100] H. Nemori, Y. Matsuda, S. Mitsuika, M. Matsui, O. Yamamoto, Y. Takeda, N. Imanishi: Stability of garnet-type solid electrolyte $Li_xLa_3A_{2-y}B_yO_{12}$ (A = Nb or Ta, B = Sc or Zr), *Solid State Ionics* 282 (2015) 7–12.

[101] L. Truong, M. Howard, O. Clemens, K. S. Knight, P. R. Slater, V. Thangadurai: Facile proton conduction in H^+/Li^+ ion-exchanged garnet-type fast Li-ion conducting $Li_5La_3Nb_2O_{12}$, *J. Mater. Chem. A* 1 (2013) 13469–13475.

[102] C. Galven, J. Dittmer, E. Suard, F. Le Berre, M. P. Crosnier-Lopez: Instability of lithium garnets against moisture. Structural characterization and dynamics of $Li_{7-x}H_xLa_3Sn_2O_{12}$ and $Li_{5-x}H_xLa_3Nb_2O_{12}$, *Chem. Mater.* 24 (2012) 3335–3345.

[103] M. Nakayama, M. Kotobuki, H. Munakata, M. Nogami, K. Kanamura: Firs-principles density functional calculation of electrochemical stability of fast Li-ion conducting garnet-type oxides, *Phys. Chem. Chem. Phys.* 14 (2012) 10008–10014.

[104] M. Kotobuki, H. Munakata, K. Kanamura, Y. Sato, T. Yoshida: Compatibility of $Li_7La_3Zr_2O_{12}$ solid electrolyte to all-solid-state battery using Li metal anode, *J. Electrochem. Soc.* 157(10) (2010) A1076–A1079.

[105] M. Kotobuki, K. Kanamura: Fabrication of all-solid-state battery using $Li_5La_3Ta_2O_{12}$ ceramic electrolyte, *Ceram. Intl.* 39 (2013) 6481–6487.

[106] S. Ohta, T. Kobayashi, J. Seki, T. Asaoka: Electrochemical performance of an all-solid-state lithium ion battery with garnet-type oxide electrolyte, *J. Power Sources* 202 (2012) 332–335.

[107] S. Ohta, J. Seki, Y. Yagi, Y. Kihira, T. Tani, T. Asaoka: Co-sinterable lithium garnet-type oxide electrolyte with cathode for all-solid-state lithium ion battery, *J. Power Sources* 265 (2014) 40–44.

[108] S. Hasegawa, N. Imanishi, T. Zhang, J. Xie, A. Hirano, Y. Takeda, O. Yamamoto: Study on lithium/air secondary batteries-Stability of NASICON-type lithium ion conducting glass-ceramics with water, *J. Power Sources* 189 (2009) 371–377.

[109] T. Zhang, N. Imanishi, S. Hasegawa, A. Hirano, J. Xie, Y. Takeda, O. Yamamoto, N. Sammes: Water-stable lithium anode with the three-layer construction for aqueous lithium-air secondary batteries, *Electrochem. Solid State Lett.* 12(7) (2009) A132–A135.

[110] T. Zhang, S. Liu, N. Imanishi, S. Hasegawa, A. Hirano, Y. Takeda, O. Yamamoto: Water-stable lithium electrode and its application in aqueous lithium/air secondary batteries, *Electrochemistry* 78(5) (2010) 360–362.

[111] T. Zhang, N. Imanishi, S. Hasegawa, A. Hirano, J. Xie, Y. Takeda, O. Yamamoto, N. Sammes: Li/Polymer electrolyte/water stable lithium-conducting glass ceramics composite for lithium-air secondary batteries with aqueous electrolyte, *J. Electrochem. Soc.* 155 (2008) A965–A969.

[112] T. Zhang, N. Imanishi, Y. Shimonishi, A. Hirano, Y. Takeda, O. Yamamoto, N. Sammes: A novel high energy density rechargeable lithium/air battery, *Chem. Comm.* 46 (2010) 1661–1663.

[113] M. Zhang, K. Takahashi, I. Uechi, Y. Takeda, O. Yamamoto, D. Im, D.-J. Lee, B. Chi, J. Pu, J. Li, N. Imanishi: Water-stable anode with $Li_{1.4}Al_{0.4}Ge_{1.6}(PO_4)_3$-$TiO_2$ sheet prepared by tape casting method for lithium-air batteries, *J. Power Sources* 235 (2013) 117–121.

[114] D. Safanama, S. Adams: High efficiency aqueous and hybrid lithium-air batteries enabled by $Li_{1.5}Al_{0.5}Ge_{1.5}(PO_4)_3$ ceramic anode-protecting membrane, *J. Power Sources* 340 (2017) 294–301.

[115] B. Yan, Y. Zhu, F. Pan, J. Liu, L. Lu: $Li_{1.5}Al_{0.5}Ge_{1.5}(PO_4)_3$ Li-ion conductor prepared by melt-quench and low temperature pressing, *Solid State Ionics* 278 (2015) 65–68.

[116] Y. Suzuki, K. Watanabe, S. Sakuma, N. Imanishi: Electrochemical performance of an all-solid-state lithium-oxygen battery under humidified oxygen, *Solid State Ionics* 289 (2016) 72–76.

[117] K. M. Abraham, Z. Jiang, B. Carroll: Highly conductive PEO-like polymer electrolytes, *Chem. Mater.* 9 (1997) 1978.

[118] B. Kumar, J. Kumar, R. Leese, J. P. Fellner, S. J. Rodrigues, K. M. Abraham: A solid-state, rechargeable, long cycle life lithium-air battery, *J. Electrochem. Soc.* 157(1) (2010) A50–A54.

[119] Y. Suzuki, K. Watanabe, S. Sakuma, N. Imanishi: Electrochemical performance of an all-solid-state lithium-oxygen battery under humidified oxygen, *Solid State Ionics* 289 (2016) 72–76.

Chapter 8

Summary

All-solid-state Li-ion battery and Li-air battery are expected as the next generation of energy storage devices. To successfully fabricate all-solid-state battery and Li-air battery, Li-ion conductors should possess high ionic conductivity, electrochemical stability and wide stable potential range. Although there are many Li-ion conductive ceramics which are capable of being used as a solid electrolyte for the all-solid-state battery, all-solid-state battery with Li-ion conductive ceramics has not yet appeared on the market, except LiPON-based thin-film battery. The reason for that is thought to be lack of knowledge of interface between electrode and electrolyte and of the ceramic process technique. The property of the interface is critical to determine the performance of the all-solid-state battery; however, this is still not clear enough. The properties of the interface need to be clarified in detail. For that, SEM observation with EDS mapping [1–3] and electrochemical impedance technique [4] and other such techniques have been used (Fig. 8.1). Li atoms are not sensitive to EDS mapping, and even with these techniques it is difficult to detect the interface at the atomic scale. These restrictions of the characterization techniques must be solved. Recently, spatial-resolved electron energy-loss spectroscopy (EELS) combined with TEM was used to characterize the electrode/electrolyte interface at the atomic scale [5, 6] (Fig. 8.2).

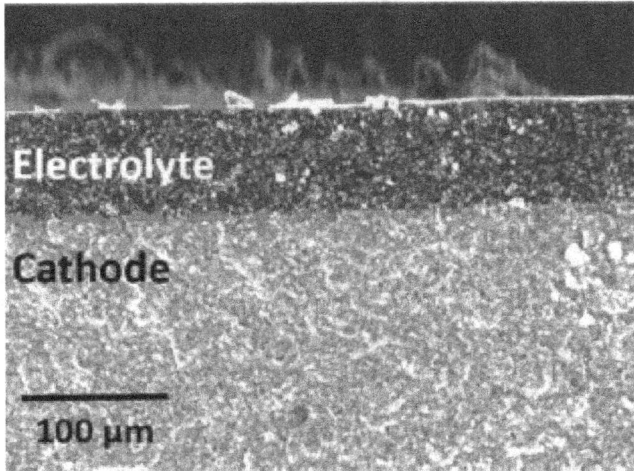

Fig. 8.1 SEM observation of electrode/electrolyte interface [1]. *Note*: Scale is micrometer level.

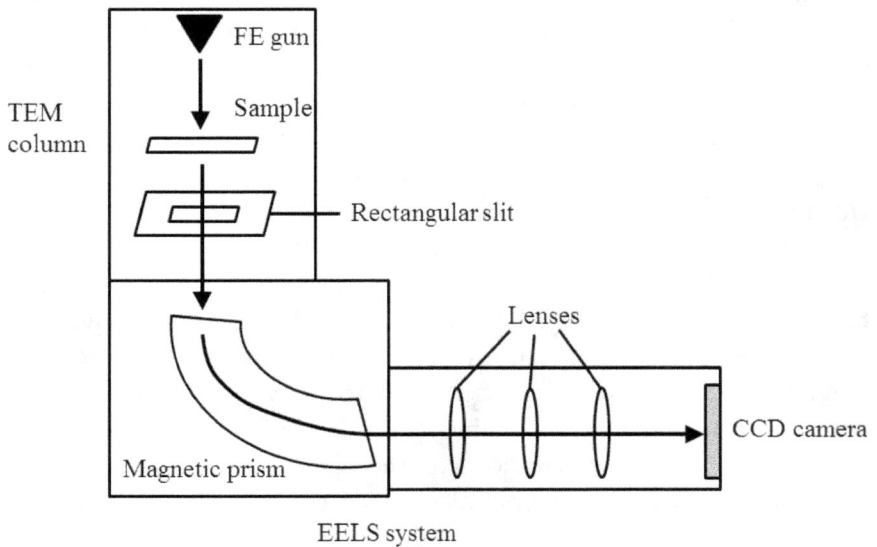

Fig. 8.2 Schematic image of spatial-resolved EELS combined with TEM.

In this system, a parallel electron beam is uniformly irradiated on the sample. The TEM image can be obtained at the bottom of the TEM column. A region of interest is selected by adjusting the rectangular slit in front of the EELS system. The energy-loss electrons due to inelastic scattering in the sample are dispersed in a magnetic prism. The electrons are magnified by some lenses and recorded by a charge-coupled-device (CCD) camera. By using this system, analysis of the interface at the atomic scale is possible. Kato *et al.* characterized $LiCo_{1/3}Ni_{1/3}Mn_{1/3}O_2/LATP$ interface and found that Co accumulation occurred around the interface [6].

Additionally, the performance of the all-solid-state battery is largely affected by the configuration of the cathode, anode and electrolyte. Some promising configurations have been suggested [7, 8] (Fig. 8.3); however, these configurations are hard to fabricate with the currently available ceramics process technique. For example, in the honeycomb type all-solid-state battery mentioned in Section 7.2, hole size and wall thickness of the honeycomb wall were $180 \times 180 \times 180\,\mu$m and ca. $100\,\mu$m, respectively. The battery performance should be improved due to the short Li-ion diffusion

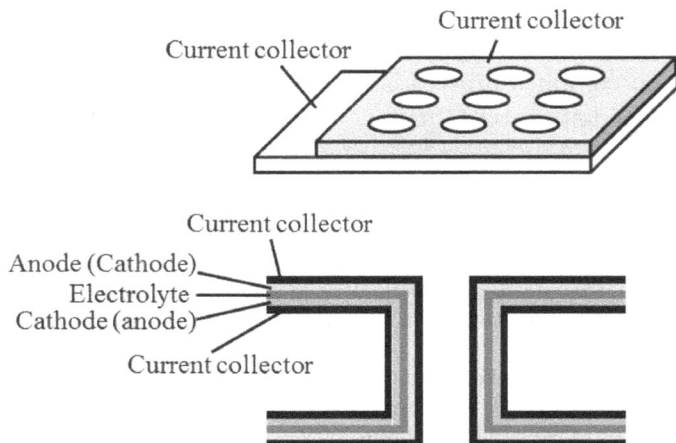

Fig. 8.3 Schematic image of 3D thin film battery based on microchannel plates [8].

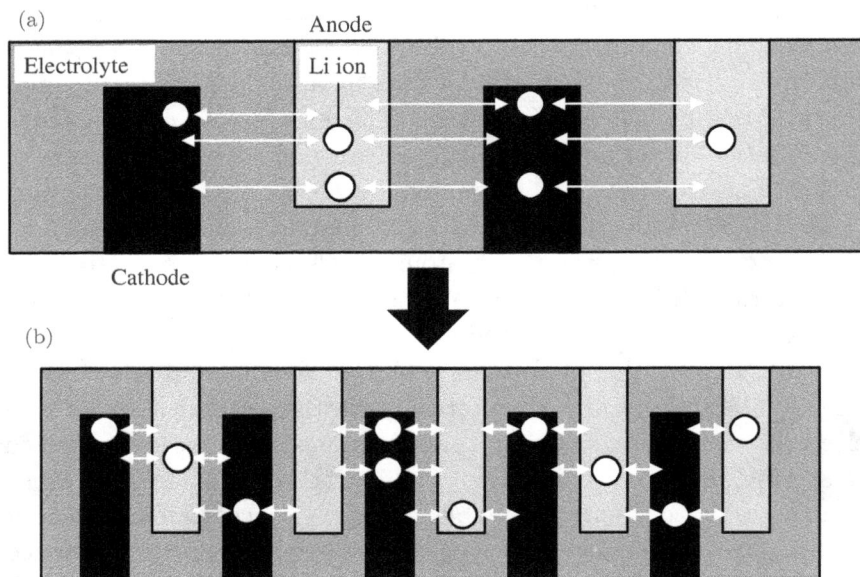

Fig. 8.4　Schematic images of honeycomb battery. (a) Thick wall and large holes, (b) thin wall and small holes.

distance if a honeycomb with smaller hole size and thinner wall would be used (Fig. 8.4). Therefore, not only material research but also research on ceramics process is intensively required for development of the all-solid-state battery.

References

[1] R.-J. Chen, Y.-B. Zhang, T. Liu, B.-Q. Xu, Y.-H. Lin, C.-W. Nan, Y. Sheng: Addressing the interface issues in all-solid-state bulk-type lithium ion battery via an all-composite approach, *ACS Appl. Mater. Interfaces* 9 (2017) 9654–9661.

[2] M. Kotobuki, Y. Suzuki, H. Munakata, K. Kanamura, Y. Sato, K. Yamamoto, T. Yoshida: Effect of sol composition on solid electrode/solid electrolyte interface for all-solid-state lithium ion battery, *Electrochim. Acta* 56 (2011) 1023–1029.

[3] T. Liu, Y. Ren, Y. Shen, S.-X. Zhao, Y. Lin, C.-W. Nan: Achieving high capacity in bulk-type solid-state lithium ion battery based on $Li_{6.75}La_3Zr_{1.75}Ta_{0.25}O_{12}$ electrolyte: Interfacial resistance, *J. Power Sources* 324 (2016) 349–357.

[4] M. Kotobuki, Y. Suzuki, H. Munakata, K. Kanamura, Y. Sato, K. Yamamoto, T. Yoshida: Fabrication of three-dimensional battery using ceramic electrolyte with honeycomb structure by sol-gel process, *J. Electrochem. Soc.* 157(4) (2010) A493–A498.

[5] K. Yamamoto, R. Yoshida, T. Sato, H. Matsumoto, H, Kurobe, T. Hamanaka, T. Kato, Y. Iriyama, T. Hirayama: Nano-scale simultaneous observation of Li-concentration profile and Ti-, O electronic structure changes in an all-solid-state Li-ion battery by spatially-resolved electron energy-loss spectroscopy *J. Power Sources* 266 (2014) 414–421.

[6] T. Kato, R. Yoshida, K. Yamamoto, T. Hirayama, M. Motoyama, W. C. West, Y. Iriyama: Effect of sintering temperatur on interfacial structure and interfacial resistance for all-solid-state rechargeable lithium batteries, *J. Power Sources* 325 (2016) 584–590.

[7] Y. Wang, B. Liu, Q. Li, S. Cartmell, S. Ferrara, Z. D. Deng, J. Xiao: Lithium and lithium ion batteries for applications in microelectronic devices: A review, *J. Power Sources* 286 (2015) 330–345.

[8] J. F. Oudenhaven, L. Baggetto, P. H. L. Notten: All-solid-state lithium-ion microbatteries: A review of various three-dimensional concepts, *Adv. Energy Mater.* 1 (2011) 10–33.

Index